OLD-FASHIONED WAY TO DESIGN A BEAM-TRUSS STEEL STRUCTURE

ANTONIO SHOPOV

2025

OLD-FASHIONED WAY TO DESIGN A BEAM-TRUSS STEEL STRUCTURE

Author: Antonio Shopov, PhD

Editor: Assoc. Prof. Radi Ganev, PhD

English, First edition, Sofia 2025

CONTENTS

1. INTRODUCTION

The task aims to design the supporting steel structure of a hangar for the Sofia area, with dimensions of 36x42 meters in plan. In essence, this is an industrial building with a multifunctional purpose, according to the needs of the technological and production part.

The development was carried out based on the following parameters of the construction site:

- Location – Sofia, Ovcha Kupel district.
- Terrain features – flat terrain;
- Climatic characteristics – temperate climate;
- Standard snow load for the region – Sn = 0.50 kN/m
- Standard wind load for the region – Wm = 0.38 kN/m
- Seismic coefficient for earthquake – Kc = 0.27
- Bearing capacity of the ground (soil) – 0.020 kN/cm^2
- Self-weight of the enclosing structure – 0.50 kN/m
- Self-weight of insulation – 0.20 kN/m^2

The geometric characteristics on building are:

- Length – Ld = 36 meters
- Width – Lw = 42 meters
- Height – H = 12 meters

A spatial structural and compositional solution was chosen, representing a beam-truss. The arrangement of the trusses in both planes forms a spatial scheme that provides opportunities for optimizing the options - both in terms of structural loads and in terms of the material capacity of the structure, respectively its applicability. The roof structure consists of a roofing sheet with an insulating coating, capitals, and the main supporting structure is a beam-truss. The loads from the beam truss are transmitted to the foundation through columns with a composite center distance of B = 6 meters in plan and a trapezoidal portal truss, located in the plane along the "x – z" axes. The spatial stability of the columns is carried out through "X" connections at both ends of the building, dimensioned solely for tension and allowing for real savings in material.

The calculations and strength are presented for the maximum loaded elements of the structure - in this case for the central beam-trusses, with two symmetrically located along the "x" axis, but due to uniformity, the calculation of only one of them is presented, and one along the "y" axis. The calculations were performed based on the existing regulatory framework (to 2000 year) - for the first and second limit states, and a check for deformations was performed. The calculations for the beam-truss and the portal structure are performed with the classic theory of structural mechanical. The main compositional elements of the dimensioned structure are: as an element base –

plane, but as a compositional production – spatial, which provides an opportunity for variation, complexity and modulation of different in their production essence and final dimensions according to the technical and technological requirements of the presented type of production building. The presented model is optimized in shape and size of the sections, under the condition of optimal material consumption to achieve minimal internal stresses and maximum reliability, with polyvariance of the implementation.

By basic structural design, the rod structures are planes, and structurally – the rod fragments are with two straight belts. The structural design of each system node is presented fragmented in the details. The portal truss is trapezoidal, with an initial height H = 2.60 meters and a ridge height H = 4.20 meters, which requires the addition of a "superstructure" with H = 1.60 m to the ridge elevation to the truss, in order to level the roof structure, to obtain a gable roof with a slope i = 5%, which is not treated in the development.

The development presents the assembly plans of the structural solution, all static and dimensional calculations of the elements of the load-bearing structure, structural drawings of the dimensioned elements of the structure, and an exemplary technology for the assembly of the structure. The building plan allows the use of the full width of the building for the implementation of the assembly processes, with a requirement for the construction of temporary roads and the tracing of all axes - in both directions of the building. When installing the columns and trusses along the "x" axis, the crane moves inside the building, and when installing those along the "y" axis outside it. The selection of the crane was carried out through a comparable correspondence of the main installation and technical parameters of the installation elements and the technical and operational characteristics of the crane. In addition to the explanatory note, the requirements for "Technique, protection and occupational safety" are presented, as the specific requirements for implementation are also given additional instructions regarding the method, marked on the drawings in the form of notes. The specific requirements for welding seams are given for the relevant elements and details.

In the implementation of the structure, steel was used - ВСт3сп4 (in EU - S235, S235J0, S235J2G3, S235JR, S235JRG2; in USA - K01804, K02001, K02301, K02502, K02601, K02701, K02702, M1017; in Japan - SS330, SS34, SS400; in China - Q235, Q235A, Q235A-B, Q235A-Z, Q235B, Q235B-Z, Q235C; in Sweden - 1312, 1313; in Finland - FORM300H, RACOLD03F, RACOLD215S; in Austria - RSt360B) according to BDS 2592-71 and electrodes type E 46. Only Bulgarian materials were used in the development.

In the presented development, an attempt was made to achieve optimality according to the main qualitative criteria (maximum) and quantitative criteria (minimum). Based on these premises, maximum functionality is actually presented, with possibilities for multivariance of implementation, reliability and reconstructibility. In the sphere of the minimum of quantitative criteria, materials, labor, time for construction of the structure, energy consumption during construction are placed, which are actually within the scope of technological criteria. Approaching the minimum of criteria from the position that the material is materialized labor, energy is transformed material, labor is a replaceable energy, each of which has a second measure of time and a third measure of value, with the last two being measures and indicators of the quantity and intensity of material, labor and energy, the optimal structure of the minimum quantity of each of the listed and its maximum intensity has been determined.

As can be seen from the last decades, many buildings have completely changed their functional purpose, which is usually expressed either in very expensive and inefficient reconstruction or through corresponding "technological inconveniences" for users. The advantage of metal structures over reinforced concrete ones and especially over prestressed reinforced concrete structures is undeniable in the conditions of "non-destructive modernizations and reconstructions". The presented development, with its constructive solution, assumes real variability both of the internal distribution and for possible subsequent expansion (reconstruction and modernization), under one single condition – the presence of terrain possibility – in horizontal or vertical plan. In this aspect, the seemingly increased cost, against the proposed subsequent options, is minimally economical and actually represents a future investment that has no real competitor in time. The adaptability of the structural cross-sections, expressed in "adaptability" to changes in the limit shear forces in the direction of the form-forming of the structural schemes, is inherent to the maximum of steel structures and is actually their indisputable advantage, which must find its maximum applicability in reality. On the other hand, the proposed structural solution is actually technologically inefficient, i.e. heterogeneous in its nature technological actions are required, associated with a high degree of control during implementation and synchronicity of incompatible contractors, since most of the installation of the main skeleton is defined as forced, although compared to the technological efficiency of reinforced concrete buildings, the presented structure is technologically many times more efficient as a final result.

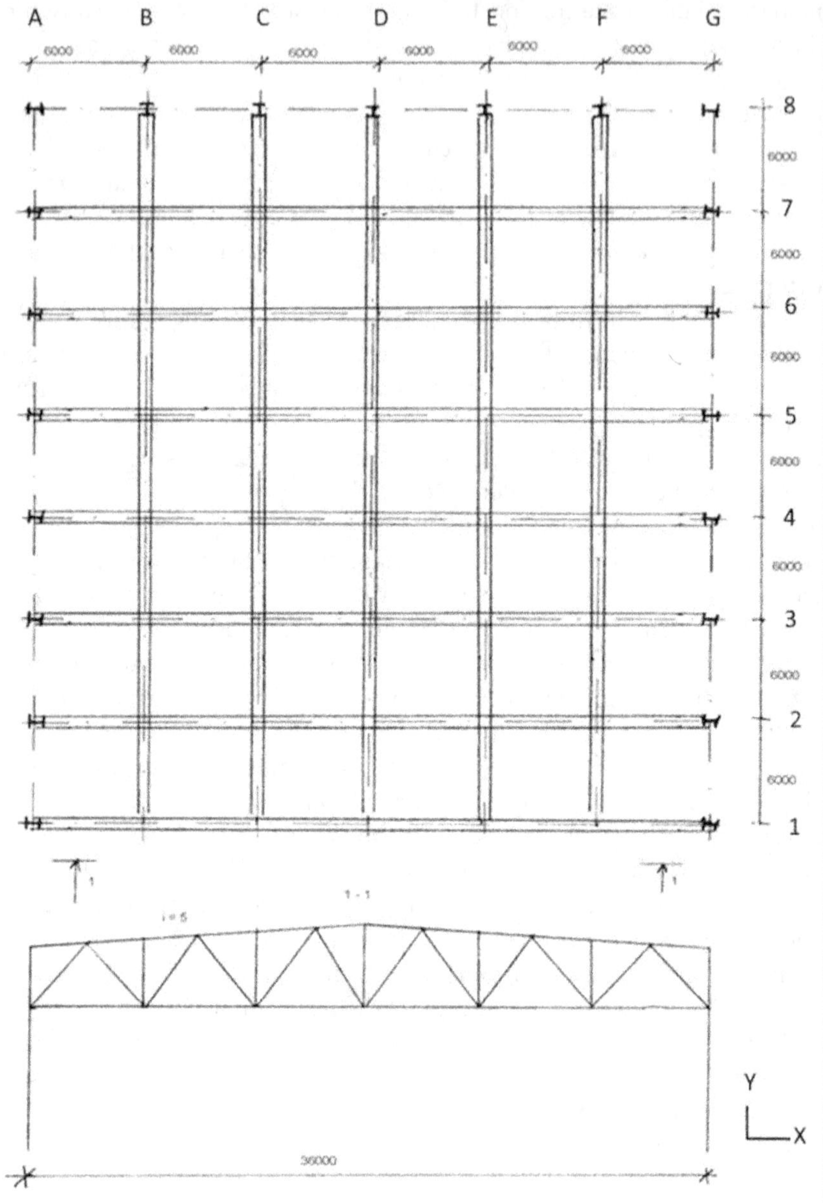

Figure 1.1 Basic scheme – all dimension in [mm]

2. ROOFING SHEET METAL

2.1. STATIC SCHEME – simple beam with a= 300 cm

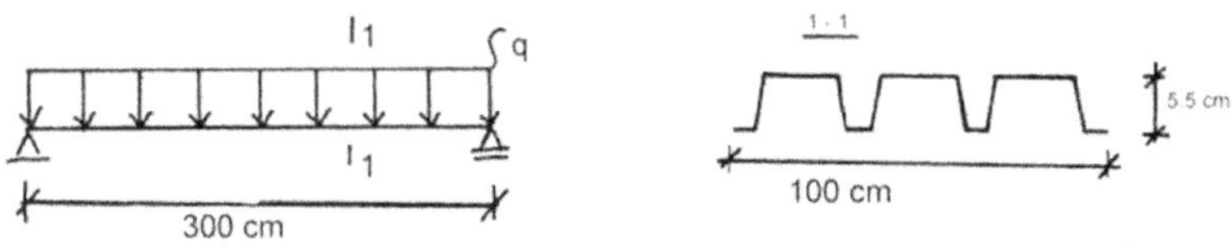

Figure 2.1 Static Scheme – simple beam

2.2. LOAD

► insulation $g_n = 0.2$ kN/m2

► specific weight of sheet metal – $g_L = 0.095$ kN/m²

$g_i = q_1 = 0.095 + 0.2 = 0.295 \approx 0.3$ kN/m² – coating weight (sheathing)

$q_2 = g_n * 1.2 + g_L * 1.05 = 0.2*1.2 + 0.095*1.05 = 0.34$ kN/m²

► snow

$s_n = s_t*\mu = 0.5*1 = 0.5$ kN/m²

$q_n = (q_1 + 0.9*s_n)*1 = (0.3 + 0.9*0.5)*1 = 0.75$ kN/m

$q = (q_2 + 0.9*s_n)*1 = (0.34 + 0.9*0.765)*1 = 1.03$ kN/m

$$\frac{g_n}{s_n} = \frac{0.34}{0.5} = 0.68 < 1 \rightarrow \gamma_{of,s} \approx 1.53 \quad \text{i.e.}$$

$s = s_n * \gamma_{f,s} = 0.5 * 1.53 = 0.765\ \mathrm{kN/m^2}$

$q_n = (q_1 + 0.9*s_n)*1 = (0.3 + 0.9*0.5)*1 = 0.75$ kN/m

$q = (q_2 + 0.9*s)*1 = (0.34 + 0.9*0.765)*1 = 1.03$ kN/m

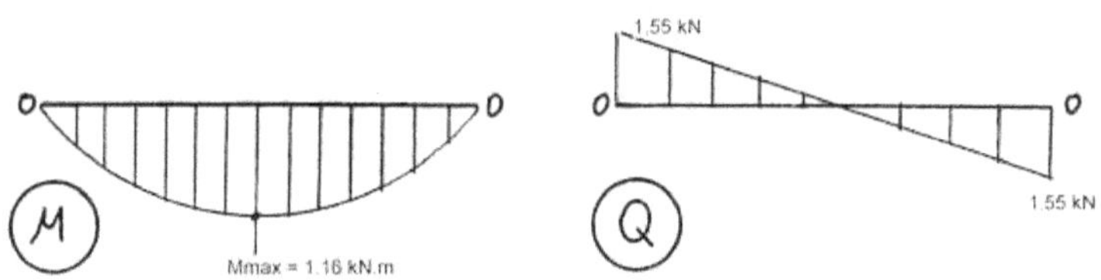

Figure 2.2 Free-body diagrams – simple beam

$M_{max} = 0.125*q*a^2 = 0.125*1.03*3^2 = 1.16$ kN.m

min $W_a = 12.40$ cm³/m – sheet metal with t = 0.8 mm and h = 55 mm,

$I_a = 45.6$ cm⁴/m

2.3. STRESS CHECK (strength check) deflection

$$\sigma = \frac{M_{max}}{\min W_a} = \frac{1.16 * 10^{-3}}{12.40 * 10^{-6}} = 9.35\ \mathrm{kN/cm^2} < 20.5\ \mathrm{kN/cm^2}$$

2.4. DEFLECTION CHECK (deformation check)

$$f = \frac{5 * q_n * a^4}{384 * E * I_a} = \frac{5 * 0.75 * 3^4 * 10^{-3}}{384 * 2.06 * 10^5 * 45.6 * 10^{-8}} = 0.00842 = 0.842\ \mathrm{cm}$$

$$f = 0.842\ \mathrm{cm} < f_u = \frac{1}{200} * a = \frac{300}{200} = 1.50\ \mathrm{cm}$$

3. CAPITALS

3.1. STATIC SCHEME – simple beam with B = 6 m, a_{ct} = 3 m, angel $\alpha = 5^0$

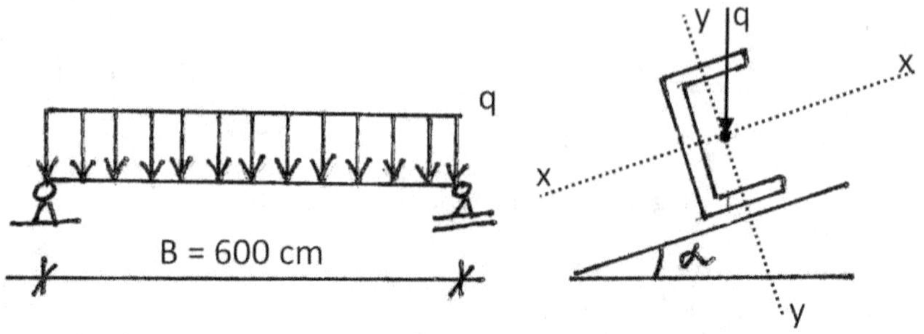

Figure 3.1 Static Scheme – simple beam

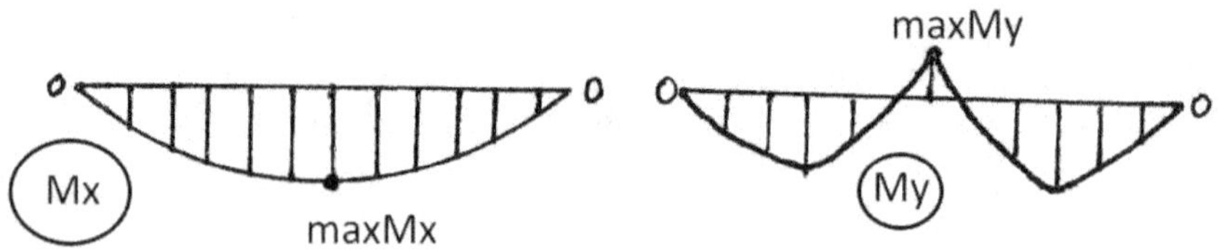

Figure 3.2 Free-body diagrams – continuous beam

3.2. LOAD

► net weight of the casing:

$$\frac{g}{\cos\alpha} * a_{ct} = \frac{0.3 * 3}{\cos 5^0} = 0.90 \text{ kN/m}$$

► capital self-weight: $g_{ct} * a_{ct} = 0{,}15*3 = 0{,}45$ kN/m

$\Sigma g_i = g_n = 0.9+0.45 = 1.35$ kN/m

$g_1 = g_n * f_{f,g} = 1.35*1.1 = 1.485$ kN/m

► snow:

$s_n = s_t * a_{ct} = 0.5*3 = 1.50$ kN/m

$\frac{g_n}{s_n} = \frac{1.35}{1.5} = 0.9 < 1 \rightarrow f_{f,s} = 1.45$

$s = s_n * f_{f,s} = 1.5*1.45 = 2.18$ kN/m

$q = g_1 + s = 1.485 + 2.18 = 3.67$ kN/m

$q_n = g_n + s_n = 1.35+1.5 = 1.85$ kN/m

$q_x = q*\cos\alpha = 3.67*\cos5^0 = 3.65$ kN/m

$q_y = q*\sin\alpha = 3.67*\sin5^0 = 0.32$ kN/m

$q_x^n = q_n*\cos\alpha = 1.85*\cos5^0 = 1.84$ kN/m

$q_y^n = q_n*\sin\alpha = 1.85*\sin5^0 = 0.16$ kN/m

$M_x = 0.125*q_x*B^2 = 0.125*3.65*6^2 = 16.425$ kN.m

$M_y = 0.0313*q_y*B^2 = 0.0313*0.32*6^2 = 0.361$ kN.m

3.3. DESIGN

Let's place hangers – one in each field.

$\sigma = \frac{M_x}{W_x} + \frac{M_y}{W_y} \leq \gamma_c * R_y$- the check that must be performed (strength)

Profile 16 according to BDS 6176-75 was selected with the following cross-section characteristics:

h= 160 mm; b= 64 mm; t_w = 5 mm; t_f = 8.4 mm; R = 8.5 mm; A = 18.1 cm^2
I_x = 747 cm^4; W_x = 93.4 cm^3; i_x = 6.42 cm; S_x = 54.1 cm; I_y = 63.3 cm^4;
W_y = 13.8 cm^3; i_y = 1.87 cm; z = 1.80 cm; I_t = 3.97 cm^4; figure 3.3

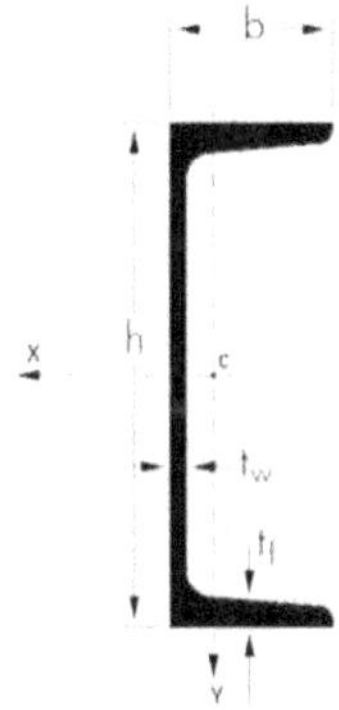

Figure 3.3 Steel profile

$$\sigma = \frac{M_x}{W_x} + \frac{M_y}{W_y} = \frac{16.425*100}{93.4} + \frac{0.361 * 100}{13.8} = 17.59 + 2.62 = 20.21\ kN/cm^2$$

$$\sigma = 20.21\ kN/cm^2 \leq \gamma_c * R_y = 21.50\ kN/cm^2$$

the strength check is completed, a check for deflection follows. (check according to the 2nd group of limit states – deformation)

$$f_x = \frac{5 * q_x{}^n * B^4}{384 * E * I_x} = \frac{5*1.84*6^4 * 10^{-3}}{384*2.06*10^5 * 747 * 10^{-8}} = 0.0202\ m = 2.02\ cm$$

$$f_y = \frac{1 * q_y{}^n * (\frac{B}{2})^4}{384 * E * I_y} = \frac{1*0.16*(\frac{6}{2})^4 * 10^{-3}}{384 * 2.06 * 10^5 * 63.3 * 10^{-8}} = 0.0030\ m = 0.30\ cm$$

$$\sqrt{f_x{}^2 + f_y{}^2} = f = \sqrt{2.02^2 + 0.30^2} = 2.04\ cm < f_u = \frac{B}{200} = \frac{600}{200} = 3\ cm$$

4. BEAM - TRUSS SUPPORTING STRUCTURE

4.1. PERMANENT LOADS

► weight of coating:

$$g_o = \frac{g * f_{f,g}}{\cos \alpha} = \frac{0.3*1.15}{\cos 5^0} = 0.35\ kN/m$$

► weight of capitals:

$$\frac{g_{ct} * f_{f,g}}{a_{ct}} = \frac{0.15*1.05}{3} = 0.05 \text{ kN/m}$$

► self-weight of connections: $0.1*1.05 = 0.11$ kN/m

► self-weight truss: $0.3*1.05 = 0.315$ kN/m

$\sum g_i = 0.315 + 0.11 + 0.05 + 0.35 = 0.83$ kN/m

$$g_n = \frac{0.3}{\cos 5^o} + \frac{0.15}{3} + 0.1 + 0.3 = 0.75 \text{ kN/m}$$

4.2. TEMPORARY LOADS

$$\frac{g_n}{s_n} = \frac{0.75}{0.5} = 1.5 > 1 \rightarrow f_{f,s} = 1.4$$

► snow: $s = s_n*f_{f,g} = 0.5*1.4 = 0.70$ kN/m

► wind:

$w_n = w_m*c*\chi_z$; $w_m = w_1*f_{f,w}$; $w_{i,n} = w_m*c_i*\chi_z$; $w_1 = 0.38$ kN/m

$c_1 = - 0.072$; $c_2 = - 0.4$; $w_m = 0.38*1.4 = 0.53$ kN/m

$w_{1n} = 0.53*(-0.072)*1.155 = - 0.044$ kN/m

$w_{2n} = 0.53*(-0.4)*1.155 = - 0.25$ kN/m

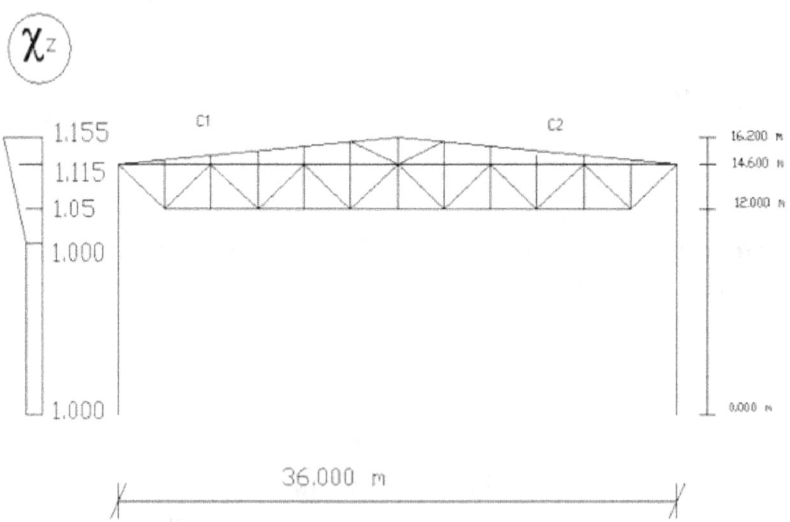

Figure 4.1 Principle for coefficient χ_z

4.3. TRANSITION TO NODAL LOADS

► computational loads

$F_g = g*a_{ct}*B = 0,83*3*6 = 14,94$ kN

$F_s = s*a_{ct}*B = 0.70*3*6 = 12.60$ kN

$F_{wi} = w_i*B*(\frac{a_{ct}}{\cos \alpha}) = w_i*6*3.01 = w_i*18.07$

F_{w1}= - 0.044*18.07 = - 0.80 kN
F_{w2}= - 0.25*18.07 = - 4.52 kN
► standard loads
F_{gn} =g_n*a_{ct}*B = 0,75*3*6 = 13,50 kN
F_{sn} = s_n*a_{ct}*B = 0.50*3*6 = 9.00 kN

4.4. PRINCIPLE OF CALCULATION ACCORDING STRUCTURAL MECHANICAL

Beam-trusses are flat, usually horizontal, frame structures loaded with loads perpendicular to their plane. They consist of intersecting beam-trusses, supported with bolts or rigidly in a supporting ring, columns or frame members. In the case of fixed support, the supporting structure must be able to absorb the transmitted bending and torsional moments.

The angle between intersecting beam-trusses can be perpendicular or the rows can intersect diagonally. It is also possible for orthogonal beam-trusses to be inserted diagonally into a rectangular or square contour. In addition to four-sided support, lattices are also available with three-sided and two-sided support. Beams parallel in both directions can have the same or different cross-sections or truss heights. Due to the rectangular or square spaces created between the intersecting beams, such structures are often called cassettes and offer very favorable architectural options for designing ceilings in spacious rooms.

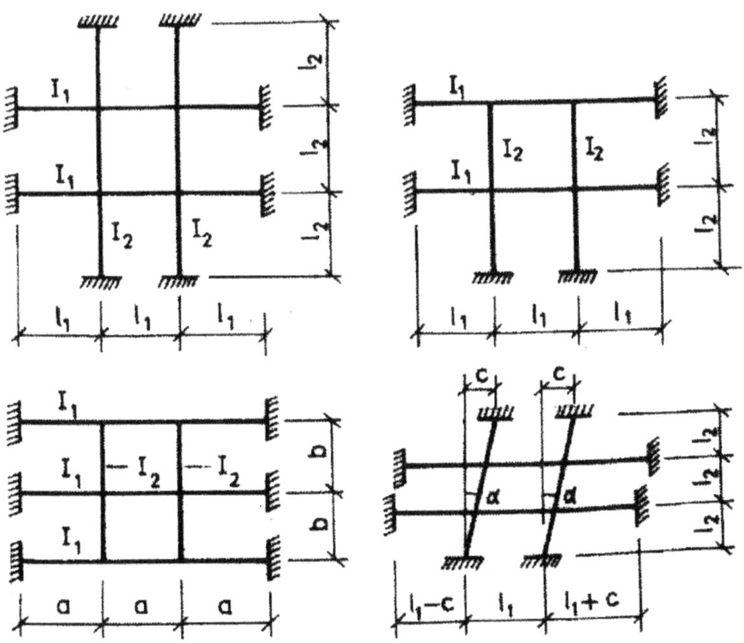

Figure 4.5 Basic main scheme beam-trusses [10]

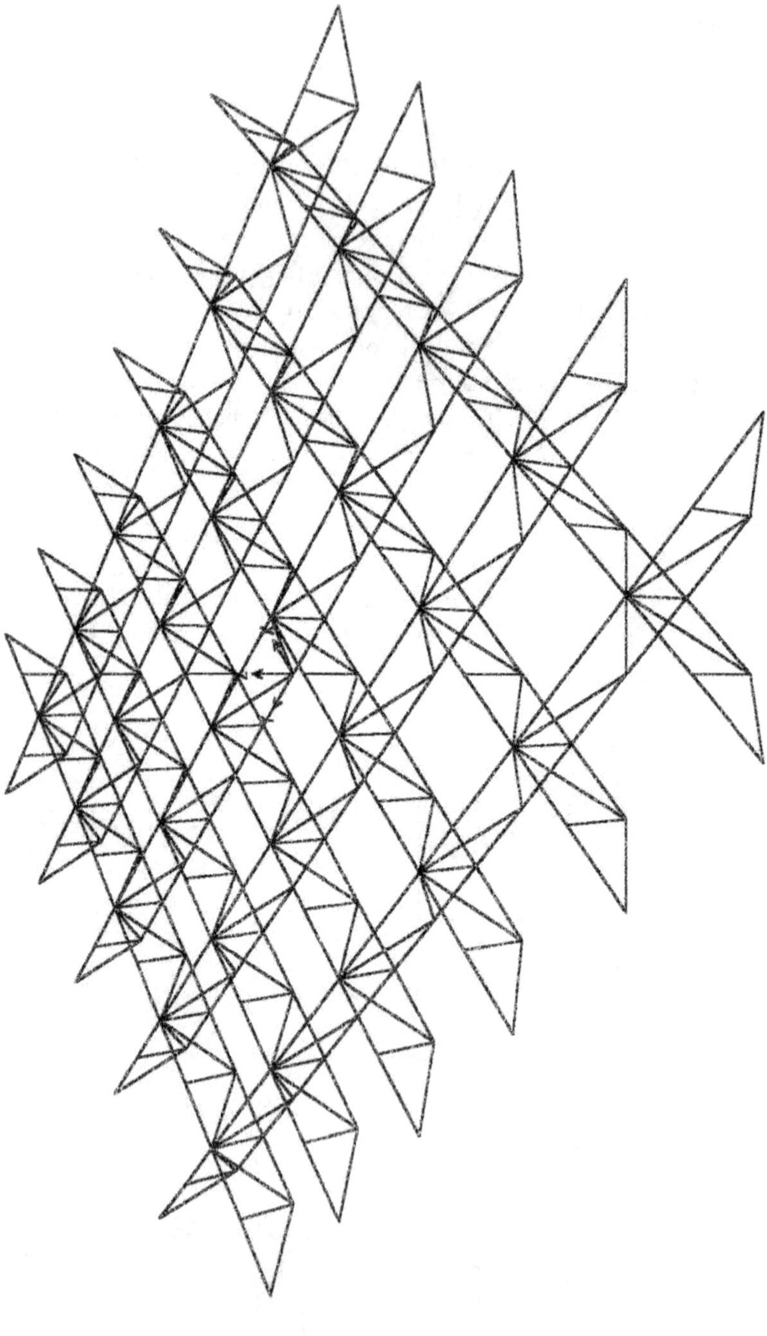

Figure 4.2 Static scheme on brem-truss structure

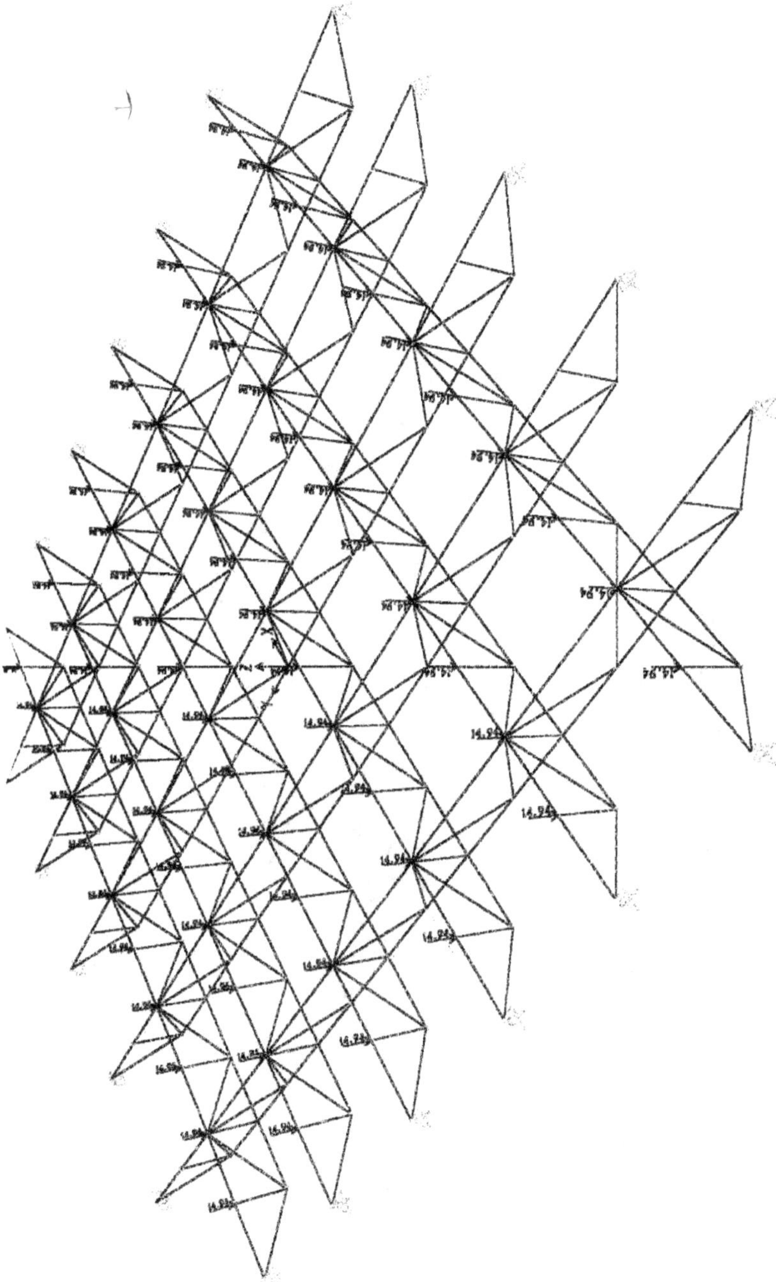

Figure 4.3 Static scheme on neam-truss structure with load

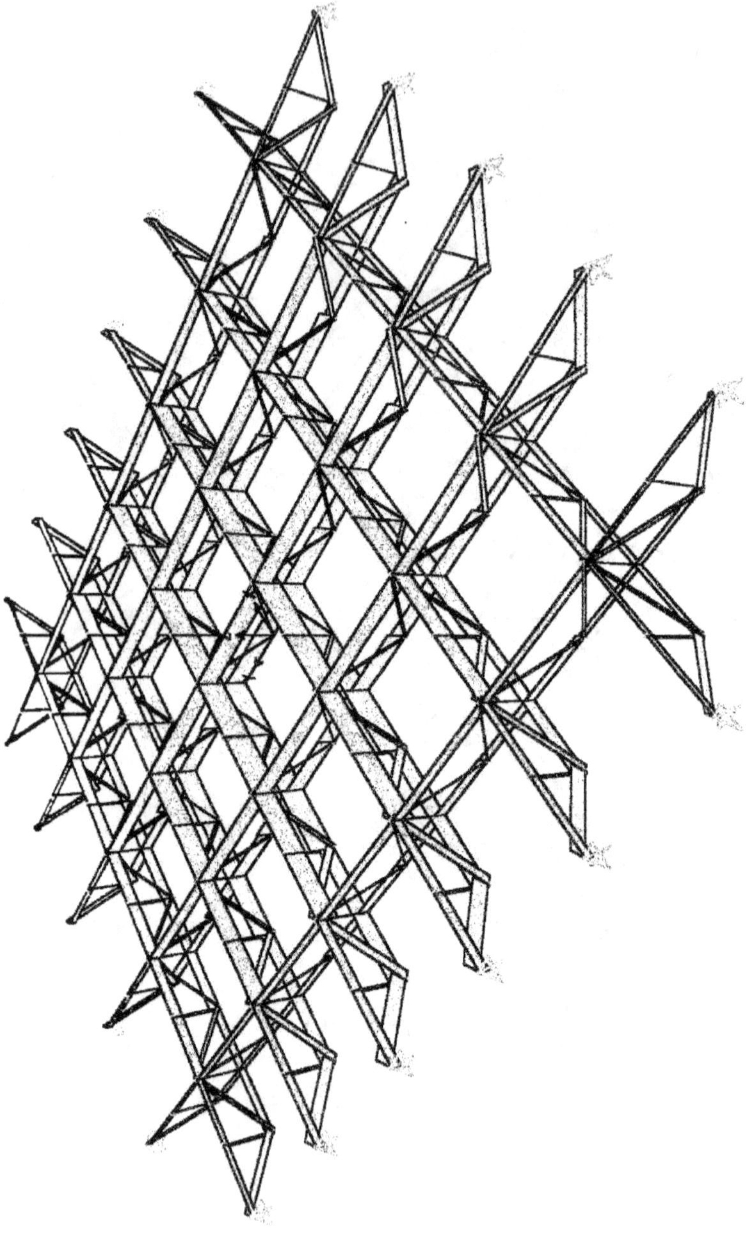

Figure 4.4 Scheme with results (one case)

5. MAIN TRUSS – PORTAL STRUCTURE

5.1. PERMANENT LOADS

► weight of coating:

$$g_o = \frac{g * f_{f,g}}{\cos \alpha} = \frac{0.3*1.15}{\cos 5^0} = 0.35 \text{ kN/m}$$

► weight of capitals:

$$\frac{g_{ct} * f_{f,g}}{a_{ct}} = \frac{0.15*1.05}{3} = 0.05 \text{ kN/m}$$

► self-weight of connections: 0.1*1.05 = 0.11 kN/m

► self-weight truss: 0.3*1.05 = 0.315 kN/m

$\sum g_i = 0.315 + 0.11 + 0.05 + 0.35 = 0.83$ kN/m

$$g_n = \frac{0.3}{\cos 5^0} + \frac{0.15}{3} + 0.1 + 0.3 = 0.75 \text{ kN/m}$$

5.2. TEMPORARY LOADS

$$\frac{g_n}{s_n} = \frac{0.75}{0.5} = 1.5 > 1 \rightarrow f_{f,s} = 1.4$$

► snow: $s = s_n * f_{f,g} = 0.5*1.4 = 0.70$ kN/m

► wind:

$w_n = w_m * c * \chi_z$; $w_m = w_1 * f_{f,w}$; $w_{i,n} = w_m * c_i * \chi_z$; $w_1 = 0.38$ kN/m

$c_1 = -0.072$; $c_2 = -0.4$; $w_m = 0.38*1.4 = 0.53$ kN/m

$w_{1n} = 0.53*(-0.072)*1.155 = -0.044$ kN/m

$w_{2n} = 0.53*(-0.4)*1.155 = -0.25$ kN/m

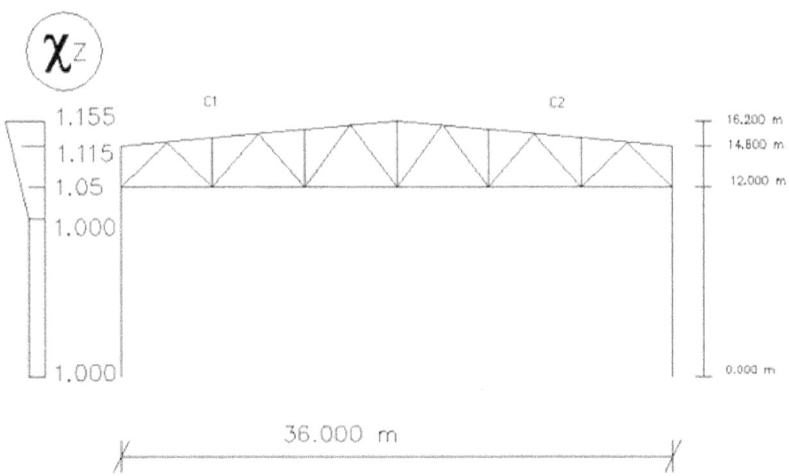

Figure 5.1 Principle for coefficient χ_z

5.3. TRANSITION TO NODAL LOADS

► computational loads

$F_g = g{*}a_{ct}{*}0.5{*}B = 0{,}83{*}3{*}0.5{*}6 = 7{,}47$ kN

$F_s = s{*}a_{ct}{*}0.5{*}B = 0.70{*}3{*}0.5{*}6 = 6.30$ kN

$F_{wi} = w_i{*}0.5{*}B{*}(\frac{a_{ct}}{\cos\alpha}) = w_i{*}0.5{*}6{*}3.01 = w_i{*}9.03$

$F_{w1} = -\ 0.044{*}9.03 = -\ 0.40$ kN

$F_{w2} = -\ 0.25{*}9.03 = -\ 2.26$ kN

► standard loads

$F_{gn} = g_n{*}a_{ct}{*}0.5{*}B = 0{,}75{*}3{*}0.5{*}6 = 6{,}75$ kN

$F_{sn} = s_n{*}a_{ct}{*}0.5{*}B = 0.50{*}3{*}0.5{*}6 = 4.50$ kN

5.4. PRINCIPLE OF CALCULATION ACCORDING STRUCTURAL MECHANICAL

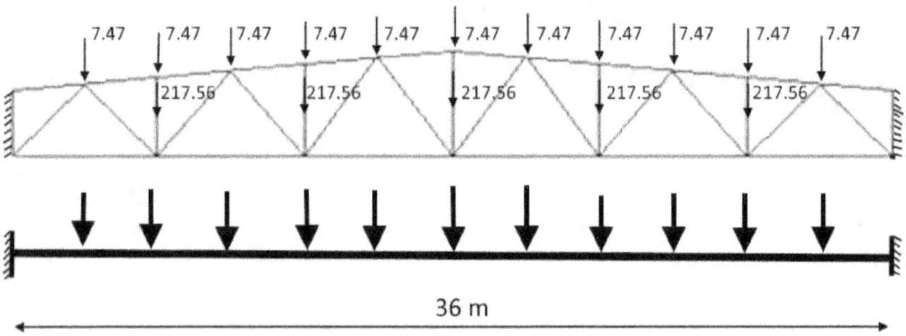

Figure 5.2 Basic main scheme on main truss

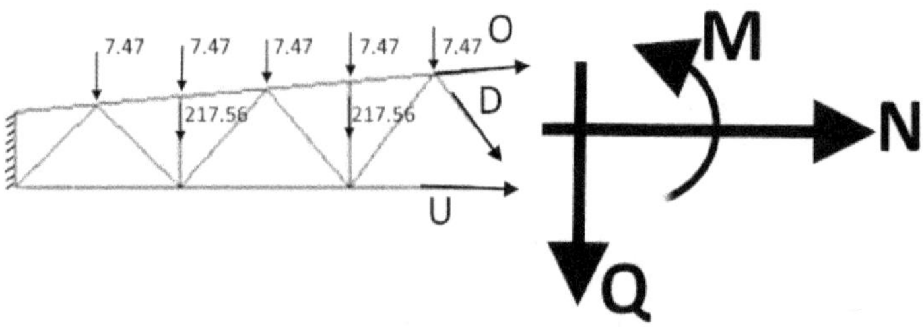

Figure 5.2 Basic main scheme for determination on force diagram

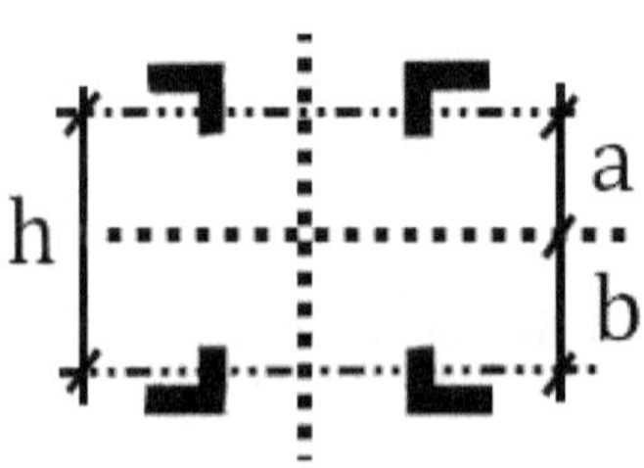

$$O = -\frac{N.b}{h} - \frac{M}{h};$$

$$U = -\frac{N.a}{h} + \frac{M}{h};$$

$$D = \frac{Q}{\sin\alpha}$$

Figure 5.3 Basic principle for determination on force diagram

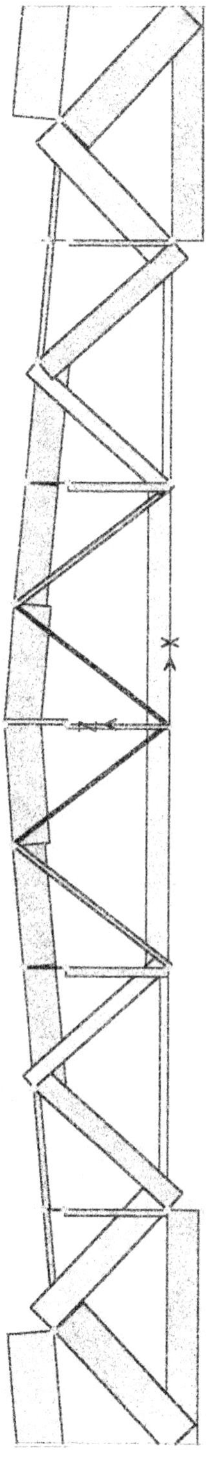

Figure 5.4 Scheme with results (one case)

6. VALUES OF FORCES IN BARS OF TRUSSES

Table 6.1 Results of calculation – main case for beam-truss on 36 m

Bars	Calculation forces in bars , kN for beam-truss on 36 m						
No.	g	S - left	S - right	S - full	W - left	W - right	Ev
01	-69,14	-50,12	-12,74	-62,45	3,79	9,44	-13,88
02	-72,10	-52,17	-13,36	-65,12	3,97	9,83	-14,47
03	-165,81	-109,41	-40,61	-149,76	10,69	20,90	-33,29
04	-167,94	-110,37	-41,57	-151,69	10,89	21,10	-33,72
05	-212,28	-115,24	-76,56	-191,74	17,34	23,05	-42,62
06	-212,97	-114,58	-77,83	-192,36	17,54	22,97	-42,76
07	-213,17	-77,83	-114,58	-192,54	22,97	17,54	-42,8
08	-212,50	-76,56	-115,24	-191,93	23,05	17,34	-42,67
09	-168,58	-41,57	-110,37	-152,27	21,10	10,89	-33,84
010	-166,46	-40,61	-109,41	-150,35	20,90	10,69	-33,42
011	-73,08	-13,36	-52,17	-66,01	9,83	3,97	-14,67
012	-70,13	-12,74	-50,12	-63,35	9,44	3,79	-14,07
U1	124,21	86,86	25,24	112,19	-7,28	-16,30	24,93
U2	127,00	88,65	26,04	114,71	-7,47	-16,66	25,49
U3	194,66	119,20	56,52	175,82	-13,90	-23,07	39,08
U4	195,72	119,19	56,60	176,78	-14,08	-23,13	39,29
U5	217,21	98,75	97,39	196,19	-20,55	-20,73	43,60
U6	217,09	97,39	97,75	196,08	-20,73	-20,55	43,58
U7	195,75	56,60	119,19	176,08	-23,13	-14,08	39,29
U8	194,44	56,52	119,20	175,63	-23,07	-13,90	39,03
U9	126,97	26,04	88,65	114,68	-16,66	-7,47	25,48
U10	124,04	25,24	86,86	112,03	-16,30	-7,28	24,89
D1	92,98	67,05	16,86	83,98	-5,15	-12,50	18,66
D2	-67,96	-45,54	-15,81	-61,38	4,28	8,63	-13,64
D3	52,79	28,16	19,45	47,68	-4,38	-5,61	10,60
D4	-34,12	-11,09	-19,7	-38,82	3,88	2,61	-6,85
D5	22,94	-4,78	25,45	20,72	-4,41	0,05	4,61
D6	-4,67	21,47	-25,65	-4,22	3,91	-3,02	-0,94
D7	-4,68	-25,65	21,47	-4,14	-3,02	3,91	-0,92
D8	22,83	25,45	-4,78	20,62	0,05	-4,41	4,59
D9	-34,04	-19,70	-11,09	-30,75	2,61	3,88	-6,84

D10	52,63	19,45	28,16	47,54	-5,64	-4,38	10,57
D11	-67,89	-15,81	-45,54	-61,32	8,63	4,28	-13,62
D12	92,83	16,86	67,05	83,85	-12,5	-5,15	18,63
V1	-13,68	-12,22	-0,12	4,40	0,41	2,18	-2,75
V2	-1,09	-0,89	-0,06	0,33	0,05	0,15	-0,22
V3	-12,82	-11,63	0,06	4,17	0,36	2,07	-2,57
V4	-0,49	-0,46	0	0,16	0,01	0,08	-0,10
V5	-12,73	-11,57	0,07	4,14	0,36	2,06	-2,56
V6	-0,47	-0,21	-0,21	0	0,04	0,04	-0,09
V7	-12,73	0,07	-11,57	1,5	2,06	0,36	-2,55
V8	-0,49	0	-0,46	0,06	0,08	0,01	-0,10
V9	-12,8	0,06	-11,63	1,51	2,07	0,36	-2,57
V10	-1,00	-0,06	-0,89	0,13	0,15	0,05	-0,2
V11	-13,65	-0,12	-12,22	1,65	2,18	0,41	-2,74
Bv	70,02	11,31	11,31	62,80	0	0	12,55
Av	70,1	51,52	51,52	62,87	0	0	12,57
Ah	1,11	0,54	0,54	1,00	0,91	0	0,22

Table 6.2 Results of calculation – combination cases for beam-truss on 36 m

Bars	Combination forces, kN					
No.	[g]+ [S left]	[g]+ [S right]	[g]+ [S full]	[g]+ [W left]	[g]+ [W right]	[g]+[s full] + 0,8*[W left]
01	-119,26	-81,88	-131,59	-65,35	-59,70	-128,56
02	-124,27	-85,46	-137,22	-68,13	-62,27	-134,04
03	-275,22	-206,42	-315,57	-155,12	-144,91	-307,02
04	-278,31	-209,51	-319,63	-157,05	-146,84	-310,92
05	-327,52	-288,84	-404,02	-194,94	-189,23	-390,15
06	-327,55	-290,8	-405,33	-195,43	-190,00	-391,3
07	-291,00	-327,75	-405,71	-190,20	-195,63	-387,33
08	-289,06	-327,74	-404,43	-189,45	-195,16	-385,99
09	-210,15	-278,95	-320,85	-147,48	-157,69	-303,97
010	-207,07	-275,87	-316,81	-145,56	-155,77	-300,09
011	-86,44	-125,25	-139,09	-63,25	-69,11	-131,23
012	-82,87	-120,25	-133,48	-60,69	-66,34	-125,93
U1	211,07	149,45	236,40	116,93	107,91	230,58
U2	215,65	153,04	241,71	119,53	110,34	235,73
U3	313,86	251,18	370,48	180,76	171,59	359,36

U4	314,91	252,32	372,50	181,64	172,59	361,24
U5	315,96	314,60	413,40	196,66	196,48	396,96
U6	314,48	314,84	413,17	196,36	196,54	396,59
U7	252,35	314,94	371,83	172,62	181,67	353,33
U8	250,96	313,64	370,07	171,37	180,54	351,61
U9	153,01	215,62	241,65	110,31	119,5	228,32
U10	149,28	210,9	236,07	107,74	116,76	223,03
D1	160,03	109,84	176,96	87,83	80,48	172,84
D2	-113,50	-83,77	-129,34	-63,68	-59,33	-125,92
D3	80,95	72,24	100,47	48,41	47,18	96,966
D4	-45,21	-53,82	-72,94	-30,24	-34,07	-69,836
D5	18,16	48,39	43,66	18,53	19,92	40,132
D6	16,80	-30,32	-8,89	-0,76	-0,76	-5,762
D7	-30,33	16,79	-8,82	-7,70	-9,09	-11,236
D8	48,28	18,05	43,45	22,88	26,71	43,49
D9	-53,74	-45,13	-64,79	-31,43	-38,42	-62,702
D10	72,08	80,79	100,17	46,99	56,91	95,658
D11	-83,70	-113,43	-129,21	-59,26	-63,61	-122,31
D12	109,69	159,88	176,68	80,33	87,68	166,68
V1	-25,90	-13,8	-9,28	-13,27	-11,5	-8,952
V2	-1,98	-1,15	-0,76	-1,04	-0,94	-0,72
V3	-24,45	-12,76	-8,65	-12,46	-10,75	-8,362
V4	-0,95	-0,49	-0,33	-0,48	-0,41	-0,322
V5	-24,3	-12,66	-8,59	-12,37	-10,67	-8,302
V6	-0,68	-0,68	-0,47	-0,43	-0,43	-0,438
V7	-12,66	-24,3	-11,23	-10,67	-12,37	-9,582
V8	-0,49	-0,95	-0,43	-0,41	-0,48	-0,366
V9	-12,74	-24,43	-11,29	-10,73	-12,44	-9,634
V10	-1,06	-1,89	-0,87	-0,85	-0,95	-0,75
V11	-13,77	-25,87	-12,00	-11,47	-13,24	-10,256
Bv	81,33	81,33	132,82	70,02	70,02	132,82
Av	121,62	121,62	132,97	70,10	70,10	132,97
Ah	1,65	1,65	2,11	2,02	1,11	2,838

Table 6.3 Results of calculation – combination cases for beam-truss on 36 m

Bars	Combination forces, kN

No.	[g] + [S full] + 0,8* [W right]	[g] + [W left] + 0,8* [S full]	[g] + [W right] + 0,8* [S full]	[g] + 0,9*[S full] +0,9* [W left]	[g] +0,9* [S full] +0,9* [W right]
O1	-124,04	-115,31	-109,66	-121,93	-116,85
O2	-129,36	-120,23	-114,37	-127,14	-121,86
O3	-298,85	-274,93	-264,72	-290,97	-281,78
O4	-302,75	-278,40	-268,19	-294,66	-285,47
O5	-385,58	-348,33	-342,62	-369,24	-364,10
O6	-386,95	-349,32	-343,89	-370,31	-365,42
O7	-391,68	-344,23	-349,66	-365,78	-370,67
O8	-390,56	-342,99	-348,70	-364,49	-369,63
O9	-312,14	-269,30	-279,51	-286,63	-295,82
O10	-308,26	-265,84	-276,05	-282,97	-292,15
O11	-135,91	-116,06	-121,92	-123,64	-128,92
O12	-130,45	-111,37	-117,02	-118,65	-123,73
U1	223,36	206,68	197,66	218,63	210,51
U2	228,38	211,30	202,11	223,52	215,25
U3	352,02	321,42	312,25	340,39	332,14
U4	354,00	323,06	314,01	342,15	334,01
U5	396,82	353,61	353,43	375,29	375,12
U6	396,73	353,22	353,40	374,91	375,07
U7	360,57	313,48	322,53	333,41	341,55
U8	358,95	311,87	321,04	331,74	340,00
U9	235,67	202,05	211,24	215,19	223,46
U10	230,25	197,36	206,38	210,20	218,32
D1	166,96	155,01	147,66	163,93	157,31
D2	-122,44	-112,78	-108,43	-119,35	-115,44
D3	95,982	86,554	85,324	91,76	90,653
D4	-72,90	-61,296	-65,126	-65,566	-69,013
D5	41,244	35,106	36,496	37,619	38,87
D6	-5,762	-4,136	-4,136	-4,949	-4,949
D7	-12,348	-11,012	-12,402	-11,124	-12,375
D8	46,554	39,376	43,206	41,433	44,88
D9	-68,294	-56,03	-63,02	-59,366	-65,657
D10	103,59	85,022	94,942	90,34	99,268
D11	-125,79	-108,32	-112,67	-115,31	-119,23
D12	172,56	147,41	154,76	157,05	163,66
V1	-7,536	-9,75	-7,98	-9,351	-7,758

V2	-0,64	-0,776	-0,676	-0,748	-0,658
V3	-6,994	-9,124	-7,414	-8,743	-7,204
V4	-0,266	-0,352	-0,282	-0,337	-0,274
V5	-6,942	-9,058	-7,358	-8,68	-7,15
V6	-0,438	-0,43	-0,43	-0,434	-0,434
V7	-10,942	-9,47	-11,17	-9,526	-11,056
V8	-0,422	-0,362	-0,432	-0,364	-0,427
V9	-11,002	-9,522	-11,232	-9,578	-11,117
V10	-0,83	-0,746	-0,846	-0,748	-0,838
V11	-11,672	-10,15	-11,92	-10,203	-11,796
Bv	132,82	120,26	120,26	126,54	126,54
Av	132,97	120,4	120,4	126,68	126,68
Ah	2,11	2,82	1,91	2,829	2,01

Table 6.4 Results of calculation – combination cases for beam-truss on 36 m

Bars	Combination forces, kN				
No.	[Gn]	0.9*[Gn] + [W left]	0.9*[Gn] + [W right]	[+] Extremally forces - [tension]	[-] Extremally forces - [compression]
O1	-67,33	-56,8	-51,15		-136,51
O2	-70,21	-59,22	-53,36		-142,35
O3	-161,48	-134,64	-124,43		-327,39
O4	-163,55	-136,31	-126,10		-331,60
O5	-206,75	-168,74	-163,03		-419,17
O6	-207,43	-169,14	-163,71		-420,53
O7	-207,61	-163,88	-169,31		-420,91
O8	-206,95	-163,21	-168,92		-419,58
O9	-164,15	-126,64	-136,85		-332,84
O10	-162,09	-124,98	-135,19		-328,65
O11	-71,14	-54,19	-60,05		-144,26
O12	-68,27	-52,01	-57,66		-138,45
U1	120,93	101,55	92,53	245,21	
U2	123,64	103,8	94,61	250,71	
U3	189,55	156,69	147,52	384,32	
U4	190,57	157,44	148,39	386,41	
U5	211,48	169,78	169,60	428,82	

U6	211,39	169,52	169,7	428,61	
U7	190,58	148,39	157,44	385,72	
U8	189,31	147,31	156,48	383,87	
U9	123,57	94,56	103,75	250,61	
U10	120,74	92,36	101,38	244,84	
D1	90,53	76,32	68,97	183,56	
D2	-66,17	-55,28	-50,93		-134,17
D3	51,42	41,90	40,57	104,24	
D4	-33,25	-26,04	-27,31		-75,39
D5	22,34	15,69	20,15	50,02	
D6	-4,61	-0,25	-7,17		-30,72
D7	-4,48	-7,05	-0,12		-30,58
D8	22,25	20,07	15,61	49,92	
D9	-33,17	-27,25	-25,98		-67,24
D10	51,27	40,51	41,77	103,94	
D11	-66,08	-50,84	-55,19		134,01
D12	90,36	68,83	76,18	183,25	
V1	-13,32	-11,58	-9,81		-26,67
V2	-2,05	-1,79	-1,69		-3,14
V3	-12,48	-10,87	-9,16		-25,36
V4	-0,91	-0,81	-0,74		-1,46
V5	-12,39	-10,79	-9,09		-25,2
V6	-0,91	-0,78	-0,78		-1,21
V7	-12,39	-9,09	-10,79		-25,2
V8	-13,36	-11,95	-12,02		-15,16
V9	-2,05	0,23	-1,48		-13,88
V10	-0,91	-0,67	-0,77		-1,89
V11	-13,29	-9,78	-11,55		-26,84

Table 6.5 Results of calculation – main cases for beam-truss on 42 m

Bars	Calculation forces in bars , kN for beam-truss on 42 м						
No.	g	S - left	S - right	S - full	W - left	W - right	Ev
01	-46,45	-20,97	-21,67	-41,96	4,42	4,56	-9,32
02	-47,78	-21,57	-22,26	-43,15	4,52	4,68	-9,59
03	-118,13	-53,34	-53,81	-106,70	11,23	11,32	-23,71
04	-151,47	-68,39	-68,61	-136,81	14,40	14,44	-30,40
05	-160,97	-72,68	-72,68	-145,39	15,30	15,30	-32,31
06	-151,93	-68,61	-68,39	-137,23	14,44	14,40	-30,50

O7	-119,14	-53,81	-53,34	-107,61	11,32	11,23	-23,91
O8	-49,29	-22,26	-21,57	-44,52	4,68	4,52	-9,89
O9	-47,98	-21,67	-20,97	-43,33	4,56	4,42	-9,63
U1	94,43	42,64	42,50	85,29	-8,98	-8,94	18,95
U2	95,14	42,96	42,90	85,93	-9,04	-9,03	19,10
U3	143,08	64,61	64,43	129,24	-13,60	-13,56	28,72
U4	143,04	64,59	64,55	129,40	-13,60	-15,58	28,71
U5	161,74	73,03	72,92	146,09	-15,37	-15,36	32,46
U6	161,60	72,97	72,99	145,96	-15,37	-15,36	32,44
U7	161,64	72,99	72,97	146,00	-15,36	-15,37	32,41
U8	161,49	72,92	73,03	145,86	-15,36	-15,37	28,69
U9	142,92	64,55	64,59	129,09	-15,58	-13,60	28,63
U10	142,66	64,43	64,61	128,85	-13,56	-13,60	19,06
U11	94,97	42,90	42,96	85,78	-9,03	-9,04	18,89
U12	94,10	42,50	42,64	84,99	-8,94	-8,98	12,79
D1	63,71	28,77	28,65	57,55	-6,06	-6,03	-11,99
D2	-59,75	-26,98	-26,92	-53,97	5,68	5,66	6,48
D3	32,29	14,58	14,48	29,16	-3,07	-3,05	-6,23
D4	-31,06	14,02	-13,96	-28,06	2,95	2,94	2,54
D5	12,67	5,72	5,66	11,44	-1,20	-1,19	-2,42
D6	-12,07	-5,45	-5,41	-10,90	1,15	1,14	0,05
D7	0,25	0,12	0,08	0,23	-0,02	-0,02	0,04
D8	0,18	0,08	0,12	0,16	-0,02	-0,02	-2,41
D9	-11,99	-5,41	-5,45	-10,83	1,14	1,15	2,52
D10	12,55	5,66	5,72	11,33	-1,19	-1,20	-6,21
D11	-30,92	-13,96	-14,02	-27,93	2,94	2,95	6,44
D12	32,07	14,48	14,58	28,96	-3,05	-3,07	-11,96
D13	-59,59	-26,92	-26,98	-53,83	5,66	5,68	12,73
D14	63,44	28,65	28,77	57,30	-6,03	-6,06	-0,12
V1	-0,60	-0,27	-0,25	0,00	0,06	0,06	-0,18
V2	-0,88	-0,40	-0,36	-0,80	0,08	0,08	-0,09
V3	-0,46	-0,21	-0,21	-0,42	0,04	0,04	-0,09
V4	-0,47	-0,21	-0,21	-0,43	0,05	0,04	-0,09
V5	-0,46	-0,21	-0,21	-0,42	0,04	0,05	-0,09
V6	-0,45	-0,21	-0,21	-0,41	0,04	0,04	-0,09
V7	-0,80	-0,36	-0,40	-0,72	0,08	0,08	-0,16
V8	-0,56	-0,25	-0,27	-0,52	0,06	0,06	-0,11
Bv	42,23	19,07	19,07	38,14	0,00	0,00	8,43
Av	42,19	19,14	19,14	38,28	0,00	0,00	8,51

Ah	1,73	0,78	0,78	1,54	0,16	0,16	0,35

Table 6.6 Results of calculation – combination cases for beam-truss on 42 m

Bars	Combination forces, kN						
No.	[g] + [S left]	[g] + [S right]	[g] + [S full]	[g] + [W left]	[g] + [W right]	[g] + [Ev]	[g]+ [S full] +0,8* [W left]
01	-67,42	-68,12	-88,41	-42,03	-41,89	-55,77	-84,87
02	-69,35	-70,04	-90,93	-43,26	-43,10	-57,37	-87,31
03	-171,47	-171,94	-224,83	-106,9	-106,81	-141,84	-215,85
04	-219,86	-220,08	-288,28	-137,07	-137,03	-181,87	-276,76
05	-233,65	-233,65	-306,36	-145,67	-145,67	-193,28	-294,12
06	-220,54	-220,32	-289,16	-137,49	-137,53	-182,43	-277,61
07	-172,95	-172,48	-226,75	-107,82	-107,91	-143,05	-217,69
08	-71,55	-70,86	-93,81	-44,61	-44,77	-59,18	-90,06
09	-69,65	-68,95	-91,31	-43,42	-43,56	-57,61	-87,66
U1	137,07	136,93	179,72	85,45	85,49	113,38	172,54
U2	138,10	138,04	181,07	86,10	86,11	114,24	173,84
U3	207,69	207,51	272,32	129,48	129,52	171,8	261,44
U4	207,63	207,59	272,44	129,44	127,46	171,75	261,56
U5	234,77	234,66	307,83	146,37	146,38	194,20	295,53
U6	234,57	234,59	307,56	146,23	146,24	194,04	295,26
U7	234,63	234,61	307,64	146,28	146,27	194,05	295,35
U8	234,41	234,52	307,35	146,13	146,12	190,18	295,06
U9	207,47	207,51	272,01	127,34	129,32	171,55	259,55
U10	207,09	207,27	271,51	129,10	129,06	161,72	260,66
U11	137,87	137,93	180,75	85,94	85,93	113,86	173,53
U12	136,6	136,74	179,09	85,16	85,12	106,89	171,94
D1	92,48	92,36	121,26	57,65	57,68	51,72	116,41
D2	-86,73	-86,67	-113,72	-54,07	-54,09	-53,27	-109,18
D3	46,87	46,77	61,45	29,22	29,24	26,06	58,99
D4	-17,04	-45,02	-59,12	-28,11	-28,12	-28,52	-56,76
D5	18,39	18,33	24,11	11,47	11,48	10,25	23,15
D6	-17,52	-17,48	-22,97	-10,92	-10,93	-12,02	-22,05
D7	0,37	0,33	0,48	0,23	0,23	0,29	0,46
D8	0,26	0,30	0,34	0,16	0,16	-2,23	0,32
D9	-17,40	-17,44	-22,82	-10,85	-10,84	-9,47	-21,90

D10	18,21	18,27	23,88	11,36	11,35	6,34	22,92
D11	-44,88	-44,94	-58,85	-27,98	-27,97	-24,48	-56,49
D12	46,55	46,65	61,03	29,02	29,00	20,11	58,59
D13	-86,51	-86,57	-113,42	-53,93	-53,91	-46,86	-108,89
D14	92,09	92,21	120,74	57,41	57,38	63,32	115,92
V1	-0,87	-0,85	-0,60	-0,54	-0,54	-0,78	-0,552
V2	-1,28	-1,24	-1,68	-0,80	-0,80	-0,97	-1,616
V3	-0,67	-0,67	-0,88	-0,42	-0,42	-0,55	-0,848
V4	-0,68	-0,68	-0,90	-0,42	-0,43	-0,56	-0,86
V5	-0,67	-0,67	-0,88	-0,42	-0,41	-0,55	-0,848
V6	-0,66	-0,66	-0,86	-0,41	-0,41	-0,54	-0,828
V7	-1,16	-1,20	-1,52	-0,72	-0,72	-0,96	-1,456
V8	-0,81	-0,83	-1,08	-0,50	-0,50	-0,67	-1,032
Bv	61,30	61,30	80,37	42,23	42,23	50,66	80,37
Av	61,33	61,33	80,47	42,19	42,19	50,7	80,47
Ah	2,51	2,51	3,27	1,89	1,89	2,08	3,398

Table 6.7 Results of calculation – combination cases for beam-truss on 42 m

Bars	Combination forces, kN					
No.	[g] + [S full] +0,8* [W right]	[g] + [W left] +0,8* [S full]	[g] + [W right] +0,8* [S full]	[g] +0,9* [S full] +0,9* [W left]	[g] +0,9* [S full] +0,9* [W right]	[g] + [Ev] + 0,8*[S] + 0,8* [W left]
O1	-84,762	-75,598	-75,458	-80,236	-80,11	-85,802
O2	-87,186	-77,78	-77,62	-82,547	-82,403	-88,274
O3	-215,77	-192,26	-192,17	-204,05	-203,97	-218,22
O4	-276,73	-246,52	-246,48	-261,64	-261,60	-279,80
O5	-294,12	-261,98	-261,98	-278,05	-278,05	-297,35
O6	-277,64	-247,27	-247,31	-262,44	-262,48	-280,66
O7	-217,77	-193,91	-194,00	-205,80	-205,88	-220,08
O8	-90,194	-80,226	-80,386	-85,146	-85,29	-91,052
O9	-87,774	-78,084	-78,224	-82,873	-82,999	-88,626
U1	172,57	153,68	153,72	163,11	163,15	174,43
U2	173,85	154,84	154,85	164,34	164,35	175,75
U3	261,47	232,87	232,91	247,16	247,19	264,31
U4	259,98	232,96	230,98	247,26	245,48	264,39
U5	295,54	263,24	263,25	279,39	279,40	298,78
U6	295,27	263,00	263,01	279,13	279,14	298,51

U7	295,34	263,08	263,07	279,22	279,21	298,56
U8	295,05	262,82	262,81	278,94	278,93	294,58
U9	261,13	230,61	232,59	245,08	246,86	262,36
U10	260,63	232,18	232,14	246,42	246,39	253,95
U11	173,52	154,56	154,55	164,05	164,04	175,26
U12	171,91	153,15	153,11	162,55	162,51	167,73
D1	116,44	103,69	103,72	110,05	110,08	92,912
D2	-109,19	-97,246	-97,266	-103,21	-103,23	-91,902
D3	59,01	52,548	52,568	55,771	55,789	46,932
D4	-56,768	-50,558	-50,568	-53,659	-53,668	-48,608
D5	23,158	20,622	20,632	21,886	21,895	18,442
D6	-22,058	-19,64	-19,65	-20,845	-20,854	-19,82
D7	0,464	0,414	0,414	0,439	0,439	0,458
D8	0,324	0,288	0,288	0,306	0,306	-2,118
D9	-21,9	-19,514	-19,504	-20,711	-20,702	-17,222
D10	22,92	20,424	20,414	21,676	21,667	14,452
D11	-56,49	-50,324	-50,314	-53,411	-53,402	-44,472
D12	58,574	52,188	52,168	55,389	55,371	40,838
D13	-108,88	-96,994	-96,974	-102,94	-102,93	-85,396
D14	115,89	103,25	103,22	109,58	109,56	104,34
V1	-0,552	-0,54	-0,54	-0,546	-0,546	-0,732
V2	-1,616	-1,44	-1,44	-1,528	-1,528	-1,546
V3	-0,848	-0,756	-0,756	-0,802	-0,802	-0,854
V4	-0,868	-0,764	-0,774	-0,812	-0,821	-0,864
V5	-0,84	-0,756	-0,746	-0,802	-0,793	-0,854
V6	-0,828	-0,738	-0,738	-0,783	-0,783	-0,836
V7	-1,456	-1,296	-1,296	-1,376	-1,376	-1,472
V8	-1,032	-0,916	-0,916	-0,974	-0,974	-1,038
Bv	80,37	72,742	72,742	76,556	76,556	81,172
Av	80,47	72,814	72,814	76,642	76,642	81,324
Ah	3,398	3,122	3,122	3,26	3,26	3,44

Table 6.8 Results of calculation – combination cases for beam-truss on 42 m

Bar	Combination forces, kN					
No.	[g] + [Ev] + 0,8*[S] + 0,8* [W right]	[Gn]	0.9*[Gn] + [W left]	0.9*[Gn] + [W right]	[+] Extremally forces - tension	[-] Extremally forces - compression

O1	-88,99	-45,23	-36,28	-36,14		-91,71
O2	-91,536	-46,52	-37,35	-37,19		-94,32
O3	-226,52	-115,01	-92,28	-92,19		-233,21
O4	-290,52	-147,47	-118,33	-118,29		-299,03
O5	-308,77	-156,72	-125,75	-125,75		-317,78
O6	-291,48	-147,93	-118,69	-118,73		-299,95
O7	-228,61	-116,00	-93,08	-93,17		-235,21
O8	-94,68	-47,99	-38,51	-38,67		-97,31
O9	-92,14	-46,71	-37,48	-37,62		-94,71
U1	181,16	91,94	73,76	73,80	186,42	
U2	182,51	92,63	74,32	74,33	187,62	
U3	274,5	139,31	111,78	111,82	282,48	
U4	272,97	139,27	111,75	109,77	282,6	
U5	310,26	157,47	126,36	126,37	319,31	
U6	309,98	157,33	126,22	126,23	319,02	
U7	310,02	157,37	126,28	126,27	319,11	
U8	306,03	157,23	126,14	126,13	318,81	
U9	274,09	139,15	109,66	111,64	282,16	
U10	364,04	138,89	111,44	111,4	281,63	
U11	181,99	92,46	74,19	74,18	187,49	
U12	174,37	91,61	73,51	73,47	185,76	
D1	97,46	62,03	49,76	49,79	125,78	
D2	-96,16	-58,17	-46,68	-46,7		-117,96
D3	49,24	31,44	25,22	25,24	63,74	
D4	-50,82	-30,24	-24,25	-24,27		-61,32
D5	19,34	12,33	9,89	9,90	25,00	
D6	-20,69	-11,75	-9,43	-9,44		-23,83
D7	0,47	0,25	0,20	0,20	0,50	
D8	-2,11	0,17	0,14	0,14		-2,22
D9	-18,06	-11,67	-9,37	-9,36		-23,67
D10	15,33	12,22	9,81	9,80	24,77	
D11	-46,66	-30,11	-24,16	-24,15		-61,05
D12	43,09	31,22	25,05	25,03	63,3	
D13	-89,61	-58,02	-46,56	-46,54		-117,65
D14	108,81	61,76	49,56	49,53	125,24	
V1	-0,77	-0,58	-0,46	-0,46		-0,91
V2	-2,48	-1,65	-1,40	-1,40		-2,61
V3	-1,39	-0,91	-0,78	-0,78		-1,42
V4	-1,40	-0,91	-0,77	-0,78		-1,43

V5	-1,37	-0,91	-0,78	-0,77		-1,42
V6	-1,37	-0,91	-0,78	-0,78		-1,41
V7	-2,48	-1,65	-1,40	-1,40		-2,53
V8	-1,12	-0,58	-0,46	-0,46		-1,16

Table 6.9 Results of calculation – main cases for main truss (portal) on 36 m

Bars	Calculation forces in bars , kN for main truss on 36 м						
No.	g	S - left	S - right	S - full	W - left	W - right	Ev
01	236,71	204,65	169,07	200,87	-34,31	-58,38	54,44
02	-28,76	-28,68	-13,51	-23,58	-1,27	14,30	-6,61
03	-32,33	-31,89	-12,23	-26,61	-0,27	14,09	-7,44
04	-163,96	-154,0	-111,68	-138,72	21,96	41,93	-37,71
05	-164,65	-154,2	-112,4	-139,32	22,18	41,88	-37,87
06	-187,70	-144,6	-144,44	-159,58	32,86	32,97	-43,17
07	-187,70	-144,4	-144,68	-159,58	32,97	32,86	-43,17
08	-164,65	-112,4	-154,24	-139,32	41,88	22,18	-37,87
09	-163,96	-111,6	-154,05	-138,72	41,93	21,96	-37,71
010	-32,33	-12,23	-31,89	-26,61	14,09	-0,27	-7,44
011	-28,76	-13,51	-28,68	-23,58	14,30	-1,27	-6,61
012	236,71	169,07	204,65	200,87	-58,38	-34,31	54,44
U1	-156,57	-127,7	-120,93	-133,64	28,72	30,62	-36,01
U2	41,57	50,28	18,42	34,32	0,62	-18,71	9,56
U3	113,74	100,67	79,32	96,19	-15,55	-25,7	26,16
U4	113,74	79,32	100,67	96,19	-25,70	-15,55	26,16
U5	41,57	18,42	50,28	34,32	-18,71	0,62	9,56
U6	-156,57	-120,9	-127,77	-133,64	30,62	28,72	-36,01
D1	-202,07	-177,2	-137,27	-170,84	24,81	55,56	-46,48
D2	164,05	144,62	114,08	138,71	- 20,76	-44,32	37,73
D3	-113,09	-104,6	-84,31	-96,25	18,77	25,46	-26,01
D4	84,19	78,27	63,10	71,76	-15,25	-16,17	19,36
D5	-25,27	3,61	-30,12	-22,03	9,56	-6,13	-5,81
D6	12,73	-11,9	22,65	11,40	-8,02	8,42	2,93
D7	12,73	22,65	-11,90	11,40	8,42	-8,02	2,93
D8	-25,27	-30,12	3,61	-22,03	-6,13	9,56	-5,81
D9	84,19	63,10	78,27	71,76	-16,17	-15,25	19,36
D10	-113,09	-84,31	-104,66	-96,25	25,46	18,77	-26,01
D11	164,05	114,08	144,62	138,71	-44,32	-20,76	37,73
D12	-202,07	-137,2	-177,21	-170,84	55,56	24,81	-46,48
V1	-30,35	-22,90	-23,96	-25,19	0,47	11,88	-6,98

V2	-40,13	-62,9	-16,55	-37,21	3,97	17,38	-10,15
V3	-24,50	-19,60	-19,60	-17,24	4,08	4,08	-5,64
V4	-44,13	-23,36	-62,90	-37,21	17,88	3,97	-10,15
V5	-30,35	-16,55	-22,90	-25,19	11,88	0,47	-6,38

Table 6.10 Results of calculation – combination cases for main truss (portal)

Bars	Combination forces, kN						
No.	[g] + [S left]	[g] + [S right]	[g] + [S full]	[g] + [W left]	[g] + [W right]	[g] + [Ev]	[g]+ [S full] +0,8* [W left]
01	441,36	405,78	437,58	202,4	178,33	291,15	410,13
02	-57,44	-42,27	-52,34	-30,03	-14,46	-35,37	-53,36
03	-64,22	-44,56	-58,94	-32,60	-18,24	-39,77	-59,16
04	-318,01	-275,64	-302,68	-142,00	-122,03	-201,67	-285,11
05	-318,89	-277,05	-303,97	-142,47	-122,77	-202,52	-286,23
06	-332,38	-332,14	-347,28	-154,84	-154,73	-230,87	-320,99
07	-332,14	-332,38	-347,28	-154,73	-154,84	-230,87	-320,90
08	-277,05	-318,89	-303,97	-122,77	-142,47	-202,52	-270,47
09	-275,64	-318,01	-302,68	-122,03	-142,00	-201,67	-269,14
010	-44,56	-64,22	-58,94	-18,24	-32,60	-39,77	-47,67
011	-42,27	-57,44	-52,34	-14,46	-30,03	-35,37	-40,90
012	405,78	441,36	437,58	178,33	202,40	291,15	390,88
U1	-284,34	-277,5	-290,21	-127,85	-125,95	-192,58	-267,23
U2	91,85	59,99	75,89	42,19	22,86	51,13	76,39
U3	214,41	193,06	209,93	98,19	88,04	139,90	197,49
U4	193,06	214,41	209,93	88,04	98,19	139,90	189,37
U5	59,99	91,85	75,89	22,86	42,19	51,13	60,92
U6	-277,5	-284,34	-290,21	-125,95	-127,85	-192,58	-265,71
D1	-379,28	-339,31	372,91	-177,26	-146,51	-248,55	-353,06
D2	308,67	278,13	302,76	143,29	119,73	201,78	286,15
D3	-217,75	-197,4	-209,34	-94,32	-87,63	-139,10	-194,32
D4	162,46	147,29	155,95	68,94	68,02	103,55	143,75
D5	-21,66	-55,39	-47,30	-15,71	-31,40	-31,08	-39,65
D6	0,83	35,38	24,13	4,71	21,15	15,66	17,71
D7	35,38	0,83	24,13	21,15	4,71	15,66	30,87
D8	-55,39	-21,66	-47,3	-31,40	-15,71	-31,08	-52,2
D9	147,29	162,46	155,95	68,02	68,94	103,55	143,01

D10	-197,40	-217,75	-209,34	-87,63	-94,32	-139,10	-188,97
D11	278,13	308,67	302,76	119,73	143,29	201,78	267,3
D12	-339,31	-379,28	-372,91	-146,51	-177,26	-248,55	-328,46
V1	-53,25	-54,31	-55,54	-29,88	-18,47	-37,33	-55,16
V2	-107,03	-60,68	-81,34	-40,16	-26,75	-54,28	-78,16
V3	-44,10	-44,10	-41,74	-20,42	-20,42	-30,14	-38,48
V4	-68,09	-107,03	-81,34	-26,75	-40,16	-54,28	-67,44
V5	-46,90	-53,25	-55,54	-18,47	-29,88	-37,33	-46,04

Table 6.11 Results of calculation – combination cases for main truss (portal)

Bars	Combination forces, kN					
No.	[g] + [S full] +0,8* [W right]	[g] + [W left] +0,8* [S full]	[g] + [W right] +0,8* [S full]	[g] +0,9* [S full] +0,9* [W left]	[g] +0,9* [S full] +0,9* [W right]	[g] + [Ev] + 0,8*[S] + 0,8* [W left]
O1	390,88	363,1	339,03	386,61	364,95	424,40
O2	-40,90	-48,89	-33,32	-51,13	-37,11	-55,25
O3	-47,67	-53,89	-39,53	-56,52	-43,60	-61,27
O4	-269,14	-252,98	-233,01	-269,04	-251,07	-295,08
O5	-270,47	-253,93	-234,23	-270,08	-252,35	-296,23
O6	-320,90	-282,50	-282,39	-301,75	-301,65	-332,25
O7	-320,99	-282,39	-282,5	-301,65	-301,75	-332,16
O8	286,23	-234,23	-253,93	-252,35	-270,08	-280,47
O9	-286,11	-233,01	-252,98	-251,07	-269,04	-279,10
O10	-59,16	-39,53	-53,89	-43,60	-56,52	-49,78
O11	-53,36	-33,32	-48,89	37,11	-51,13	-42,80
O12	410,13	339,03	363,10	364,95	386,61	405,15
U1	-265,71	-234,76	-232,86	-251,00	-249,29	-276,52
U2	60,92	69,65	50,32	73,02	55,62	79,08
U3	189,37	175,14	164,99	186,32	177,18	204,41
U4	197,49	164,99	175,14	177,18	186,32	196,29
U5	76,39	50,32	69,65	55,62	73,02	63,62
U6	-267,23	-232,96	-234,76	-249,29	-251,00	-275,00
D1	-328,46	-313,93	-283,18	-333,50	-305,82	-365,37
D2	267,30	254,26	230,70	270,21	249,00	296,14
D3	-168,97	-171,32	-164,63	-182,82	-176,80	-201,08
D4	143,01	126,35	125,43	135,05	134,22	148,76
D5	-52,20	-33,33	-49,02	-36,49	-50,61	-41,06

D6	30,87	13,83	30,27	15,77	30,57	18,36
D7	17,71	30,27	13,83	30,57	15,77	31,51
D8	-39,65	-49,02	-33,33	-50,61	-36,49	-53,61
D9	143,75	124,43	126,35	134,22	135,05	148,03
D10	-194,32	-164,63	-171,32	-176,80	-182,82	-195,73
D11	286,15	230,70	254,26	249,00	270,21	277,29
D12	-353,06	-283,18	313,93	-305,82	-333,50	-340,77
V1	-46,04	-50,03	-38,62	-52,60	-42,33	-57,11
V2	-67,44	-69,93	-6,52	-74,05	-61,98	-80,87
V3	-38,46	-34,12	-34,21	-36,34	-36,34	-40,66
V4	-78,16	-56,52	-69,93	-61,98	-74,05	-70,14
V5	-55,16	-38,62	-50,03	-42,33	-52,60	-47,98

Table 6.12 Results of calculation – combination cases for main truss (portal)

Bars	Combination forces, kN					
No.	[g] + [Ev] + 0,8*[S] + 0,8* [W right]	[Gn]	0.9*[Gn] + [W left]	0.9*[Gn] + [W right]	[+] Extremally forces [tension]	[-] Extremally forces [compression]
01	405,15	215,19	159,36	135,29	441,36	
02	-42,80	-26,15	-24,80	-9,23		-57,44
03	-49,78	-29,39	-26,72	-12,36		-64,22
04	-279,10	-149,05	-112,19	-92,22		-318,01
05	-280,47	-149,68	-112,53	-92,83		-318,89
06	-332,16	-170,64	-120,71	-120,60		-347,28
07	-332,25	-170,64	-120,6	-120,71		-347,28
08	-296,23	-149,68	-92,83	-112,53		-318,89
09	-295,08	-149,05	-92,22	-112,19		-318,01
010	-61,27	-29,39	-12,36	-26,72		-64,22
011	-55,25	-26,15	-9,23	-24,80		-57,44
012	424,40	215,19	135,29	159,36	441,36	
U1	-275,00	-142,34	-99,38	-97,48		-290,21
U2	63,62	37,79	34,63	15,3	91,85	
U3	196,29	103,40	77,51	67,36	214,41	
U4	204,41	103,40	67,36	77,51	214,41	
U5	79,08	37,79	15,30	34,63	91,85	
U6	-276,52	-142,34	-97,48	-99,36		-290,21

D1	-340,77	-183,7	-140,52	-109,77		-379,28
D2	277,29	149,14	113,46	89,90	308,67	
D3	-195,73	-102,81	-73,76	-67,07		217,75
D4	148,03	76,54	53,63	52,71	162,46	
D5	-53,61	-22,97	-11,12	-26,81		-55,39
D6	31,51	11,57	2,40	18,84	35,38	
D7	18,36	11,57	18,84	2,40	35,38	
D8	-41,06	-22,97	-26,81	-11,12		-55,39
D9	148,76	76,54	52,71	53,63	162,46	
D10	-201,08	-102,81	-67,07	-73,76		-217,75
D11	296,14	149,14	89,90	113,46	308,67	
D12	-365,37	-183,7	-109,77	-140,52		-379,28
V1	-47,98	-27,59	-24,36	-12,95		-57,11
V2	-70,14	-40,12	-32,14	-18,73		-107,03
V3	-40,66	-22,27	-15,97	-15,97		-44,1
V4	-80,87	-40,12	-18,73	-32,14		-107,03
V5	-57,11	-27,59	-12,95	-24,36		-57,11

7. CALCULATION OF BARS IN TRUSSES

7.1. BEAM - TRUSS ALONG THE "X" AXIS (WITH LENGTH L = 36 METER)

► upper bar - O_{min} = - 420.91 kN (compression)

$$A_{neobx} = \frac{O_{min}}{\phi * \gamma_c * R_y} = \frac{420.91}{0.6 * 0.8 * 22.5} = 38.97\ \text{cm}^2$$

steel profile (cross – section) ⌋ ⌊ 125.125.10

$t_{pl} = 10\ \text{mm},\ A = 48.60\ \text{cm}^2, i_x = 3.85\ \text{cm}, i_y = 5.52\ \text{cm}$

$l_{ef.x} = \mu_x * l_{geom,x} = 1 * 300 = 300$ см , $l_{ef,y} = \mu_y * l_{geom,y} = 1*600 = 600$ см

$$\lambda_x = \frac{l_{ef,x}}{i_x} = \frac{300}{3.85} = 77.92,\ \lambda_y = \frac{l_{ef,y}}{i_y} = \frac{600}{5.52} = 108.69,$$

$\lambda_{max} = 108.69 \longrightarrow \phi = 0.516$

$$\sigma = \frac{O_{min}}{\phi * A} = \frac{420.91}{48.60*0.516} = 16.78\ \text{kN/cm}^2$$

► lower bar - U_{max} = 428.82 kN (tension)

$$A_{neobx} = \frac{U_{max}}{\gamma_c * R_y} = \frac{428.82}{0.95 * 22.5}\ 20.06\ \text{cm}^2$$

steel profile (cross – section) ⌋ ⌊ 80.80.7

$t_{pl} = 10\ \text{mm},\ A = 21.60\ \text{cm}^2, i_x = 2.45\ \text{cm}, i_y = 3.67\ \text{cm}$

$l_{ef,x} = 600 \text{ cm} \ , \ l_{ef,y} = 600 \text{ cm} \ , \lambda_x = \dfrac{600}{2.45} = 244.90 \ < \ \lambda_{gr} = \ 400 \, ,$

$\lambda_y = \dfrac{600}{3.67} = 163.48 \ < \ \lambda_{gr} = 400$

$\sigma = \dfrac{U_{max}}{A} = \dfrac{428.82}{21.60} = 19.83 \text{ kN/cm}^2$

► diagonal - D_1= 183.56 kN (tension)

$A_{neobx} = \dfrac{D_1}{\gamma_c * R_y} = \dfrac{183.56}{0.95 * 22.5} = \ 8.59 \ \text{cm}^2$

steel profile (cross – section) ⌋⌊ 50.50.5

$t_{pl} = 10 \text{ mm} \, , A = \ 9.60 \text{ cm}^2 \, , i_x = 1.53 \text{ cm} \ , \ i_y = 2.45 \text{ cm}$

$l_{ef,x} = 0.8 * l_{geom,x} = 0.8 * 397 = 317.6 \text{ cm},$

$l_{ef,y} = 1 * l_{geom,y} = 1 * 397 = 397 \text{ cm}$

$\lambda_x = \dfrac{317{,}6}{1.53} = 207.58 \ < \ \lambda_{gr} = 400 \quad , \quad \lambda_y = \dfrac{397}{2.45} = 162.04 \ < \ \lambda_{gr} = 400$

$\sigma = \dfrac{D_1}{A} = \dfrac{183.56}{9.60} = 19.12 \text{ kN/cm}^2$

this profile is for all tension diagonals (D_1, D_3, D_5, D_7 and etc.).

► diagonal – D_2 = - 134.17 kN (compression)

$A_{neobx} = \dfrac{D_2}{\phi * \gamma_c * R_y} = \dfrac{134.17}{0.6 * 0.8 * 22.5} = 12.42 \text{ cm}^2$

steel profile (cross – section) ⌋⌊ 80.80.7

$t_{pl} = 10 \text{ mm}, \ A = \ 21.60 \text{ cm}^2 \, , i_x = 2.45 \text{ cm} \, , i_y = 3.67 \ \text{ cm}$

$l_{ef,x} = 317.6 \text{ cm}, l_{ef,y} = 397 \text{ cm}, \ \lambda_x = \dfrac{317.6}{2.45} = 129.63, \lambda_y = \dfrac{397}{3.67} = 108.17$

$\lambda_{max} = 129.63 \longrightarrow \phi = 0.393$

$\sigma = \dfrac{D_2}{\phi * A} = \dfrac{134.17}{0.393 * 21.60} = 15.81 \text{ kN/cm}^2$

► diagonal – D_4 = - 75.39 kN (compression)

$A_{neobx} = \dfrac{D_4}{\phi * \gamma_c * R_y} = \dfrac{75.39}{0.6 * 0.8 * 22.5} = 7.00 \text{ cm}^2$

steel profile (cross – section) ⌋⌊ 70.70.5

$t_{pl} = 10 \text{ mm} \ , \ A = \ 13.72 \text{ cm}^2 \, , i_x = 2.16 \text{ cm} \, , i_y = 3.23 \ \text{ cm}$

$l_{ef,x} = 317.6 \text{ cm}, l_{ef,y} = 397 \text{ cm}, \ \lambda_x = \dfrac{317.6}{2.16} = 147.04, \ \lambda_y = \dfrac{397}{3.23} = 122.91$

$\lambda_{max} = 147.04 \longrightarrow \phi = 0.311$

$\sigma = \dfrac{D_4}{\phi * A} = \dfrac{75.39}{0.311 * 13.72} = 17.67 \text{ kN/cm}^2$

► diagonal – D_6 = - 30.78 kN (compression)

$$A_{neobx} = \frac{D_6}{\phi * \gamma_c * R_y} = \frac{30.78}{0.6 * 0.8 * 22.5} = 2.85\ cm^2$$

steel profile (cross – section) ⅃ L 56.56.5

$t_{pl} = 10\ mm\ ,\ A = 10.82\ cm^2\ , i_x = 1.72\ cm\ , i_y = 2.69\ cm$

$$l_{ef,x} = 317.6\ cm, l_{ef,y} = 397\ cm, \lambda_x = \frac{317.6}{1.72} = 184.65, \lambda_y = \frac{397}{2.69} = 147.58$$

$\lambda_{max} = 184.65 \longrightarrow \phi = 0.203$

$$\sigma = \frac{D_6}{\phi * A} = \frac{30.78}{0.203 * 10.82} = 14.00\ kN/cm^2$$

► vertical – $V_{min/max}$ = - 26.87 kN (compression / tension)

$$A_{neobx} = \frac{V_{min/max}}{\phi * \gamma_c * R_y} = \frac{26.87}{0.6 * 0.8 * 22.5} = 2.50\ cm^2$$

steel profile (cross – section) ⅃ L 70.70.5

$t_{pl} = 10\ mm\ ,\ A = 13.72\ cm^2\ , i_\eta = 1.39\ cm\ , i_\xi = 2.72\ cm$

$$l_{ef,\eta} = 317.6\ cm, l_{ef,\xi} = 397\ cm, \lambda_\eta = \frac{260}{1.39} = 187.05, \lambda_\xi = \frac{260}{2.72} = 95.59$$

$\lambda_{max} = 187.05 \longrightarrow \phi = 0.198$

$$\sigma = \frac{D_6}{\phi * A} = \frac{26.78}{0.198 * 13.72} = 9.90\ kN/cm^2$$

this profile is for all verticals (V_1, V_2, V_3 and etc.).

7.2. BEAM - TRUSS ALONG THE "Y" AXIS (WITH LENGTH L = 42 METER)

► upper bar - O_{min} = - 317.78 kN (compression)

$$A_{neobx} = \frac{O_{min}}{\phi * \gamma_c * R_y} = \frac{317.78}{0.6 * 0.8 * 22.5} = 29.42\ cm^2$$

steel profile (cross – section) ⅃ L 140.140.10

$t_{pl} = 10\ mm,\ A = 54.60\ cm^2, i_x = 4.33\ cm, i_y = 6.12\ cm$

$l_{ef.x} = \mu_x * l_{geom,x} = 1 * 600 = 600$ см , $l_{ef,y} = \mu_y * l_{geom,y} = 1*600 = 600$ см

$$\lambda_x = \frac{l_{ef,x}}{i_x} = \frac{600}{4.33} = 138.57,\ \lambda_y = \frac{l_{ef,y}}{i_y} = \frac{600}{6.12} = 98.03,$$

$\lambda_{max} = 138.57 \longrightarrow \phi = 0.348$

$$\sigma = \frac{O_{min}}{\phi * A} = \frac{317.78}{54.60*0.348} = 16.72\ kN/cm^2$$

► lower bar - U_{max} = 319.31 kN (tension)

$$A_{neobx} = \frac{U_{max}}{\gamma_c * R_y} = \frac{319.31}{0.95 * 22.5}\ 14.93\ cm^2$$

steel profile (cross – section) ⅃ L 70.70.7

$t_{pl} = 10\text{ mm}$, $A = 18.84\text{ cm}^2$, $i_x = 2.14\text{ cm}$, $i_y = 3.28\text{ cm}$

$l_{ef,x} = 600\text{ cm}$, $l_{ef,y} = 600\text{ cm}$, $\lambda_x = \frac{600}{2.14} = 280.73 < \lambda_{gr} = 400$,

$\lambda_y = \frac{600}{3.28} = 182.92 < \lambda_{gr} = 400$

$\sigma = \frac{U_{max}}{A} = \frac{319.31}{21.60} = 16.95\text{ kN/cm}^2$

► diagonal - D_1= 125.78 kN (tension)

$A_{neobx} = \frac{D_1}{\gamma_c * R_y} = \frac{125.78}{0.95 * 22.5} = 5.88\text{ cm}^2$

steel profile (cross – section) ⅃ L 50.50.5

$t_{pl} = 10\text{ mm}$, $A = 9.60\text{ cm}^2$, $i_x = 1.53\text{ cm}$, $i_y = 2.45\text{ cm}$

$l_{ef,x} = 0.8 * l_{geom,x} = 0.8 * 397 = 317.6\text{ cm}$,

$l_{ef,y} = 1 * l_{geom,y} = 1 * 397 = 397\text{ cm}$

$\lambda_x = \frac{317,6}{1.53} = 207.58 < \lambda_{gr} = 400$, $\lambda_y = \frac{397}{2.45} = 162.04 < \lambda_{gr} = 400$

$\sigma = \frac{D_1}{A} = \frac{125.78}{9.60} = 13.10\text{ kN/cm}^2$

this profile is for all tension diagonals (D_1, D_3, D_5, D_7 and etc.).

► diagonal – D_2 = - 117.96 kN (compression)

$A_{neobx} = \frac{D_2}{\phi * \gamma_c * R_y} = \frac{117.96}{0.6 * 0.8 * 22.5} = 10.91\text{ cm}^2$

steel profile (cross – section) ⅃ L 80.80.7

$t_{pl} = 10\text{ mm}$, $A = 21.60\text{ cm}^2$, $i_x = 2.45\text{ cm}$, $i_y = 3.67\text{ cm}$

$l_{ef,x} = 317.6\text{ cm}$, $l_{ef,y} = 397\text{ cm}$, $\lambda_x = \frac{317.6}{2.45} = 129.63$, $\lambda_y = \frac{397}{3.67} = 108.17$

$\lambda_{max} = 129.63 \longrightarrow \phi = 0.393$

$\sigma = \frac{D_2}{\phi * A} = \frac{117.96}{0.393 * 21.60} = 13.90\text{ kN/cm}^2$

► diagonal – D_4 = - 61.32 kN (compression)

$A_{neobx} = \frac{D_4}{\phi * \gamma_c * R_y} = \frac{61.32}{0.6 * 0.8 * 22.5} = 5.68\text{ cm}^2$

steel profile (cross – section) ⅃ L 63.63.6

$t_{pl} = 10\text{ mm}$, $A = 14.56\text{ cm}^2$, $i_x = 1.93\text{ cm}$, $i_y = 2.99\text{ cm}$

$l_{ef,x} = 317.6\text{ cm}$, $l_{ef,y} = 397\text{ cm}$, $\lambda_x = \frac{317.6}{1.93} = 164.56$, $\lambda_y = \frac{397}{2.99} = 132.78$

$\lambda_{max} = 164.56 \longrightarrow \phi = 0.252$

$\sigma = \frac{D_4}{\phi * A} = \frac{61.32}{0.252 * 14.56} = 16.71\text{ kN/cm}^2$

► diagonal – D_6 = - 23.83 kN (compression)

$$A_{neobx} = \frac{D_6}{\phi * \gamma_c * R_y} = \frac{23.83}{0.6 * 0.8 * 22.5} = 2.21 \text{ cm}^2$$

steel profile (cross – section) ⅃ L 50.50.5

$t_{pl} = 10 \text{ mm}$, $A = 9.60 \text{ cm}^2$, $i_x = 1.53 \text{ cm}$, $i_y = 2.45 \text{ cm}$

$$l_{ef,x} = 317.6 \text{ cm}, l_{ef,y} = 397 \text{ cm}, \lambda_x = \frac{317.6}{1.53} = 207.58, \lambda_y = \frac{397}{2.45} = 162.04$$

$\lambda_{max} = 207.58 \longrightarrow \phi = 0.163$

$$\sigma = \frac{D_6}{\phi * A} = \frac{23.83}{0.163 * 9.60} = 15.23 \text{ kN/cm}^2$$

7.3. MAIN TRUSS – PORTAL STRUCTURE (WITH LENGTH L = 36 METER)

► upper bar - O_{min} = - 347.28 kN (compression); O_{max} = + 441.36 kN (tension)

$$A_{neobx} = \frac{O_{min}}{\phi * \gamma_c * R_y} = \frac{347.28}{0.6 * 0.8 * 22.5} = 32.16 \text{ cm}^2$$

steel profile (cross – section) ⅃ L 125.125.8

$t_{pl} = 10 \text{ mm}$, $A = 39.40 \text{ cm}^2$, $i_x = 3.87 \text{ cm}$, $i_y = 5.46 \text{ cm}$

$l_{ef.x} = \mu_x * l_{geom,x} = 1 * 302 = 302 \text{ cm}$, $l_{ef,y} = \mu_y * l_{geom,y} = 1*600 = 600 \text{ cm}$

$$\lambda_x = \frac{l_{ef,x}}{i_x} = \frac{302}{3.87} = 78.04, \ \lambda_y = \frac{l_{ef,y}}{i_y} = \frac{600}{5.46} = 109.89,$$

$\lambda_{max} = 109.89 \longrightarrow \phi = 0.520$

$$\sigma_{min} = \frac{O_{min}}{\phi * A} = \frac{347.28}{49.40*0.520} = 16.95 \text{ kN/cm}^2 \text{ (compression stress)}$$

$$\sigma_{max} = \frac{O_{max}}{\gamma_c * R_y} = \frac{441.36}{0.95 * 22.5} = 13.67 \text{ kN/cm}^2 \text{ (tension stress)}$$

► lower bar - U_{max} = 214.41 kN (tension); U_{min} = - 290.21 kN (compression)

$$A_{neobx} = \frac{U_{min}}{\phi * \gamma_c * R_y} = \frac{290.21}{0.6 * 0.8 * 22.5} = 22.63 \text{ cm}^2$$

steel profile (cross – section) ⅃ L 125.125.12

$t_{pl} = 10 \text{ mm}$, $A = 57.80 \text{ cm}^2$, $i_x = 3.82 \text{ cm}$, $i_y = 5.52 \text{ cm}$

$$l_{ef,x} = 600 \text{ cm} , l_{ef,y} = 600 \text{ cm} , \lambda_x = \frac{600}{3.82} = 157.08$$

$$\lambda_y = \frac{600}{5.52} = 108.70; \ \lambda_{max} = 108.70 \longrightarrow \phi = 0.277$$

$$\sigma_{min} = \frac{U_{min}}{\phi * A} = \frac{290.21}{57.80*0.277} = 18.13 \text{ kN/cm}^2 \text{ (compression stress)}$$

$$\sigma_{max} = \frac{U_{max}}{\gamma_c * R_y} = \frac{214.41}{0.95 * 57.80} = 3.90\ kN/cm^2 \quad \text{(tension stress)}$$

► diagonal – D_1 = - 379.28 kN (compression)

$$A_{neobx} = \frac{D_1}{\phi * \gamma_c * R_y} = \frac{379.28}{0.6 * 0.8 * 22.5} = 29.57\ cm^2$$

steel profile (cross – section) ⅃ L 100.100.10

$t_{pl} = 10$ mm, $A = 38.40\ cm^2$, $i_x = 3.05$ cm, $i_y = 4.52$ cm

$l_{ef.x} = \mu_x * l_{geom,x} = 0.8 * 415 = 332$ cm, $l_{ef,y} = \mu_y * l_{geom,y} = 1*415 = 415$ cm

$$\lambda_x = \frac{l_{ef,x}}{i_x} = \frac{332}{3.05} = 108.85,\ \lambda_y = \frac{l_{ef,y}}{i_y} = \frac{415}{4.52} = 91.81,$$

$\lambda_{max} = 108.85 \longrightarrow \phi = 0.512$

$$\sigma = \frac{D_1}{\phi * A} = \frac{379.28}{0.512 * 21.60} = 19.29\ kN/cm^2$$

► diagonal – D_2 = 308.67 kN (tension)

$$A_{neobx} = \frac{D_2}{\gamma_c * R_y} = \frac{308.67}{0.95 * 22.5} = 14.44\ cm^2$$

steel profile (cross – section) ⅃ L 63.63.6

$t_{pl} = 10$ mm, $A = 14.56\ cm^2$, $i_x = 1.93$ cm, $i_y = 2.99$ cm

$l_{ef,x} = 0.8 * l_{geom,x} = 0.8 * 415 = 332$ cm,

$l_{ef,y} = 1 * l_{geom,y} = 1 * 415 = 415$ cm

$$\lambda_x = \frac{332}{1.93} = 172.02 < \lambda_{gr} = 400\ ,\quad \lambda_y = \frac{415}{2.99} = 138.80 < \lambda_{gr} = 400$$

$$\sigma = \frac{D_2}{A} = \frac{308.67}{14.56} = 21.19\ kN/cm^2$$

► diagonal – D_3 = - 217.75 kN (compression)

$$A_{neobx} = \frac{D_3}{\phi * \gamma_c * R_y} = \frac{217.75}{0.6 * 0.8 * 22.5} = 20.16\ cm^2$$

steel profile (cross – section) ⅃ L 100.100.8

$t_{pl} = 10$ mm, $A = 31.20\ cm^2$, $i_x = 3.08$ cm, $i_y = 4.47$ cm

$l_{ef.x} = \mu_x * l_{geom,x} = 0.8 * 454 = 363.2$ cm, $l_{ef,y} = \mu_y * l_{geom,y} = 1*454 = 454$ cm

$$\lambda_x = \frac{l_{ef,x}}{i_x} = \frac{363.2}{3.08} = 117.92,\ \lambda_y = \frac{l_{ef,y}}{i_y} = \frac{454}{4.47} = 101.56,$$

$\lambda_{max} = 117.92 \longrightarrow \phi = 0.492$

$$\sigma = \frac{D_3}{\phi * A} = \frac{217.75}{0.492 * 21.60} = 14.18\ kN/cm^2$$

► diagonal – D_4 = 162.46 kN (tension)

$$A_{neobx} = \frac{D_4}{\gamma_c * R_y} = \frac{162.46}{0.95 * 22.5} = 7.60 \ \ cm^2$$

steel profile (cross – section) ⌋⌊ 63.63.6

$t_{pl} = 10$ mm , $A = 14.56 \ cm^2$, $i_x = 1.93$ cm , $i_y = 2.99$ cm

$l_{ef,x} = 0.8 * l_{geom,x} = 0.8 * 454 = 363.2$ cm,

$l_{ef,y} = 1 * l_{geom,y} = 1 * 454 = 454$ cm

$$\lambda_x = \frac{363.2}{1.93} = 188.19 < \lambda_{gr} = 400 \ , \quad \lambda_y = \frac{454}{2.99} = 151.84 < \lambda_{gr} = 400$$

$$\sigma = \frac{D_4}{A} = \frac{162.46}{14.56} = 11.16 \ kN/cm^2$$

► diagonal – D_5 = - 55.39 kN (compression)

$$A_{neobx} = \frac{D_5}{\phi * \gamma_c * R_y} = \frac{55.39}{0.6 * 0.8 * 22.5} = 5.12 \ cm^2$$

steel profile (cross – section) ⌋⌊ 70.70.6

$t_{pl} = 10$ mm, $A = 16.30 \ cm^2$, $i_x = 2.15$ cm , $i_y = 3.25$ cm

$l_{ef.x} = \mu_x * l_{geom,x} = 0.8 * 496 = 396.8$ cm, $l_{ef,y} = \mu_y * l_{geom,y} = 1*496 = 496$ cm

$$\lambda_x = \frac{l_{ef,x}}{i_x} = \frac{396.8}{2.15} = 184.56, \ \lambda_y = \frac{l_{ef,y}}{i_y} = \frac{496}{4.25} = 152.62,$$

$\lambda_{max} = 184.56 \longrightarrow \phi = 0.203$

$$\sigma = \frac{D_5}{\phi * A} = \frac{55.39}{0.203 * 21.60} = 16.74 \ kN/cm^2$$

► diagonal – D_6 = 35.38 kN (tension)

$$A_{neobx} = \frac{D_6}{\gamma_c * R_y} = \frac{35.38}{0.95 * 22.5} = 1.66 \ \ cm^2$$

steel profile (cross – section) ⌋⌊ 70.70.6

$t_{pl} = 10$ mm , $A = 16.30 \ cm^2$, $i_x = 2.15$ cm , $i_y = 3.25$ cm

$l_{ef,x} = 0.8 * l_{geom,x} = 0.8 * 496 = 396.8$ cm,

$l_{ef,y} = 1 * l_{geom,y} = 1 * 496 = 496$ cm

$$\lambda_x = \frac{396.8}{2.15} = 184.56 < \lambda_{gr} = 400 \ , \quad \lambda_y = \frac{496}{3.25} = 152.62 < \lambda_{gr} = 400$$

$$\sigma = \frac{D_6}{A} = \frac{35.38}{16.30} = 2.17 \ kN/cm^2$$

► vertical – V_1 = - 57.11 kN (compression)

$$A_{neobx} = \frac{V_1}{\phi * \gamma_c * R_y} = \frac{57.11}{0.6 * 0.8 * 22.5} = 5.28 \ cm^2$$

steel profile (cross – section) ⌋⌊ 90.90.8

$t_{pl} = 10$ mm , $A = 21.20 \ cm^2$, $i_\eta = 1.79$ cm , $i_\xi = 3.50$ cm

$l_{ef,\eta} = 313 \text{ cm}, l_{ef,\xi} = 313 \text{ cm}, \lambda_\eta = \frac{313}{1.79} = 174.86, \lambda_\xi = \frac{313}{3.50} = 89.43$

$\lambda_{max} = 174.86 \longrightarrow \phi = 0.224$

$\sigma = \frac{V_1}{\phi * A} = \frac{57.11}{0.224 * 21.20} = 12.03 \text{ kN/cm}^2$

► vertical – V_2 = - 107.03 kN (compression)

$A_{neobx} = \frac{V_2}{\phi * \gamma_c * R_y} = \frac{107.03}{0.6 * 0.8 * 22.5} = 9.91 \text{ cm}^2$

steel profile (cross – section) ⌋ ⌊ 100.100.8

$t_{pl} = 10 \text{ mm} \ , \ A = 31.20 \text{ cm}^2 \ , i_\eta = 1.98 \text{ cm} \ , i_\xi = 3.88 \text{ cm}$

$l_{ef,\eta} = 367 \text{ cm}, l_{ef,\xi} = 367 \text{ cm}, \lambda_\eta = \frac{367}{1.98} = 185.35, \lambda_\xi = \frac{367}{3.88} = 94.59$

$\lambda_{max} = 185.35 \longrightarrow \phi = 0.201$

$\sigma = \frac{V_2}{\phi * A} = \frac{107.03}{0.201 * 31.20} = 17.07 \text{ kN/cm}^2$

► vertical – V_3 = - 44.10 kN (compression)

$A_{neobx} = \frac{V_3}{\phi * \gamma_c * R_y} = \frac{44.10}{0.6 * 0.8 * 22.5} = 4.08 \text{ cm}^2$

steel profile (cross – section) ⌋ ⌊ 125.125.8

$t_{pl} = 10 \text{ mm} \ , \ A = 39.40 \text{ cm}^2 \ , i_\eta = 2.49 \text{ cm} \ , i_\xi = 4.87 \text{ cm}$

$l_{ef,\eta} = 420 \text{ cm}, l_{ef,\xi} = 420 \text{ cm}, \lambda_\eta = \frac{420}{2.49} = 168.68, \lambda_\xi = \frac{420}{4.87} = 86.77$

$\lambda_{max} = 168.68 \longrightarrow \phi = 0.236$

$\sigma = \frac{V_3}{\phi * A} = \frac{44.10}{0.236 * 39.40} = 4.74 \text{ kN/cm}^2$

► period (vibration) – T_1 = 0.27 sec (first period)

► max deflection (max deformation) on beam-trusses steel structure

$f = 11.60 \text{ cm} < f_u = \frac{L}{250} = \frac{3600}{250} = 14.40 \text{ cm}$

► max deflection (max deformation) on main truss (portal structure)

$f = 9.70 \text{ cm} < f_u = \frac{L}{250} = \frac{3600}{250} = 14.40 \text{ cm}$

8. CROSS-SECTION OF BARS IN TRUSSES

Table 8.1 Results of calculation in beam-trusses

bars	extremally forces		section profile	
No.	+ [tension]	- [compression]	type	dimension
Ox max		-420,91	⌋ ⌊	125.125.10
Ux max	428,82		⌋ ⌊	80.80.8
Dx1	183,56		⌋ ⌊	50.50.5
Dx2		-134,17	⌋ ⌊	80.80.7
Dx3	104,24		⌋ ⌊	50.50.5
Dx4		-75,39	⌋ ⌊	70.70.5
Dx5	50,02		⌋ ⌊	50.50.5
Dx6		-30,78	⌋ ⌊	56.56.5
Oy max		-317,78	⌋ ⌊	140.140.10
Uy max	319,31		⌋ ⌊	70.70.7
Dy1	125,78		⌋ ⌊	50.50.5
Dy2		-117,96	⌋ ⌊	80.80.7
Dy3	63,74		⌋ ⌊	50.50.5
Dy4		-61,32	⌋ ⌊	63.63.6
Dy5	25,00		⌋ ⌊	50.50.5
Dy6		-23,83	⌋ ⌊	50.50.5
Dy7	0,50		⌋ ⌊	50.50.5
Vx1		-26,87	⌋ ⌊	70.70.5
Vx2		-3,14	⌋ ⌊	70.70.5
Vx3		-25,36	⌋ ⌊	70.70.5
Vx4		-1,46	⌋ ⌊	70.70.5
Vx5		-25,20	⌋ ⌊	70.70.5
Vx6		-1,21	⌋ ⌊	70.70.5
Vy1		-0,91	⌋ ⌊	50.50.5
Vy2		-2,61	⌋ ⌊	50.50.5
Vy3		-1,42	⌋ ⌊	50.50.5
Vy4		-1,43	⌋ ⌊	50.50.5

Table 8.2 Results of calculation in beam-trusses

Bars	Area	Lx	Ly	ix	iy	λ_{max}	ϕ, min
No.	[cm2]	[cm]	[cm]	[cm]	[cm]		
Ox max	48,60	300	600	3,85	5,44	110,29	0,52
Ux max	24,60	600	600	2,44	3,62	245,90	
Dx1	9,60	317,6	397	1,53	2,38	207,58	
Dx2	21,60	317,6	397	2,45	3,60	129,63	0,41
Dx3	9,60	317,6	397	1,53	2,38	207,58	
Dx4	13,72	317,6	397	2,16	3,15	147,04	0,326
Dx5	9,60	317,6	397	1,53	2,38	207,58	
Dx6	9,92	317,6	397	1,72	2,69	162,87	0,211
Oy max	54,6	600	600	4,33	6,12	138,57	0,348
Uy max	16,30	600	600	2,15	3,18	279,07	
Dy1	9,60	317,6	397	1,53	2,38	207,58	
Dy2	21,60	317,6	397	2,44	3,60	129,63	0,408
Dy3	9,60	317,6	397	1,53	2,38	207,58	
Dy4	14,56	317,6	397	1,93	2,91	164,56	0,262
Dy5	9,60	317,6	397	1,53	2,38	207,58	
Dy6	9,60	317,6	397	1,53	2,38	207,58	0,17
Dy7	9,60	317,6	397	1,53	2,38	207,58	
Vx1	13,72	260	260	1,39	2,72	187,05	0,206
Vx2	13,72	260	260	1,39	2,72	187,05	0,206
Vx3	13,72	260	260	1,39	2,72	187,05	0,206
Vx4	13,72	260	260	1,39	2,72	187,05	0,206
Vx5	13,72	260	260	1,39	2,72	187,05	0,206
Vx6	13,72	260	260	1,39	2,72	187,05	0,206
Vy1	9,60	260	260	1,92	0,98	163,27	0,266
Vy2	9,60	260	260	1,92	0,98	163,27	0,266
Vy3	9,60	260	260	1,92	0,98	163,27	0,266
Vy4	9,60	260	260	1,92	0,98	163,27	0,266

Table 8.3 Results of calculation of welding rod in beam-trusses

	forces	welding rod		Calculated - length of welding rod		Accepted - length of welding rod	
bars	N	kf1	kf2	Lw1	Lw2	Lw1	Lw2
No.	[kN]	[cm]	[cm]	[cm]	[cm]	[cm]	[cm]

Dx1	183,56	0,4	0,4	15,16	7,07	16,00	8,00
Dx2	134,17	0,4	0,4	11,35	5,44	12,00	6,00
Dx3	104,24	0,4	0,4	9,04	4,45	10,00	5,00
Dx4	75,39	0,4	0,4	6,82	3,49	7,00	5,00
Dx5	50,02	0,4	0,4	4,86	2,65	5,00	5,00
Dx6	30,78	0,4	0,4	3,38	2,02	5,00	5,00
Dy1	125,78	0,4	0,4	10,71	5,16	11,00	6,00
Dy2	117,96	0,4	0,4	10,10	4,90	11,00	5,00
Dy3	63,74	0,4	0,4	5,92	3,11	6,00	5,00
Dy4	61,32	0,4	0,4	5,73	3,03	6,00	5,00
Dy5	25,00	0,4	0,4	2,93	1,83	5,00	5,00
Dy6	23,83	0,4	0,4	2,84	1,79	5,00	5,00
Dy7	0,50	0,4	0,4	1,04	1,02	5,00	5,00
Vx1	26,87	0,4	0,4	3,07	1,89	5,00	5,00
Vx2	3,14	0,4	0,4	1,24	1,10	5,00	5,00
Vx3	25,36	0,4	0,4	2,96	1,84	5,00	5,00
Vx4	1,46	0,4	0,4	1,11	1,05	5,00	5,00
Vx5	25,20	0,4	0,4	2,94	1,83	5,00	5,00
Vx6	1,21	0,4	0,4	1,09	1,04	5,00	5,00
Vy1	0,91	0,4	0,4	1,07	1,03	5,00	5,00
Vy2	2,61	0,4	0,4	1,20	1,09	5,00	5,00
Vy3	1,42	0,4	0,4	1,11	1,05	5,00	5,00
Vy4	1,43	0,4	0,4	1,11	1,05	5,00	5,00

Table 8.4 Results of calculation in main-truss

bars	extremally forces		section profile	
No.	+ [tension]	- [compression]	type	dimension
O max	441,36	-347,28	⅃ L	125.125.8
U max	214,41	-290,21	⅃ L	125.125.12
D1		-379,28	⅃ L	100.100.10
D2	308,60		⅃ L	63.63.6
D3		-271,75	⅃ L	100.100.8
D4	162,46		⅃ L	63.63.6
D5		-55,39	⅃ L	70.70.6
D6	35,38		⅃ L	70.70.6
V1		-57,11	⅃ L	90.90.8
V2		-107,03	⅃ L	100.100.8
V3		-44,10	⅃ L	125.125.8

Table 8.5 Results of calculation in main-truss

Bars	Area	Lx	Ly	ix	iy	λ_{max}	ϕ, min
No.	[cm2]	[cm]	[cm]	[cm]	[cm]		
O max	39,40	302	600	3,68	5,46	109,89	0,52
U max	57,80	600	600	3,62	5,52	157,08	0,277
D1	38,40	332	415	3,05	4,52	108,85	0,512
D2	14,56	332	415	1,93	2,99		
D3	31,20	363,2	454	3,06	4,47	117,92	0,492
D4	14,56	363,2	454	1,93	2,99		
D5	16,30	369,8	496	2,15	3,25	184,56	0,203
D6	16,32	369,8	496	2,15	3,25		
V1	21,20	313	313	1,79	3,50	174,86	0,234
V2	31,20	367	367	3,07	4,47	185,35	0,211
V3	39,40	420	420	4,67	2,49	168,86	0,236

Table 8.6 Results of calculation of welding rod in beam-trusses

	forces	welding rod		Calculated - length of welding rod		Accepted - length of welding rod	
bars	N	kf1	kf2	Lw1	Lw2	Lw1	Lw2
No.	[kN]	[cm]	[cm]	[cm]	[cm]	[cm]	[cm]
D1	379,28	0,5	0,5	17,39	8,02	18,00	9,00
D2	308,60	0,5	0,5	14,33	6,71	15,00	7,00
D3	217,75	0,5	0,5	9,82	5,03	10,00	5,00
D4	162,46	0,5	0,5	8,02	4,01	8,00	5,00
D5	55,39	0,5	0,5	3,39	2,03	5,00	5,00
D6	35,36	0,5	0,5	2,53	1,66	5,00	5,00
V1	57,11	0,5	0,5	2,76	2,76	5,00	5,00
V2	107,03	0,5	0,5	4,30	4,30	5,00	5,00
V3	44,10	0,5	0,5	2,36	2,36	5,00	5,00

9. COLUMN AND FUNDATION ON "4" AXIS BY "X"

9.1. STATIC SCHEME AND LOADS – figure 9.1

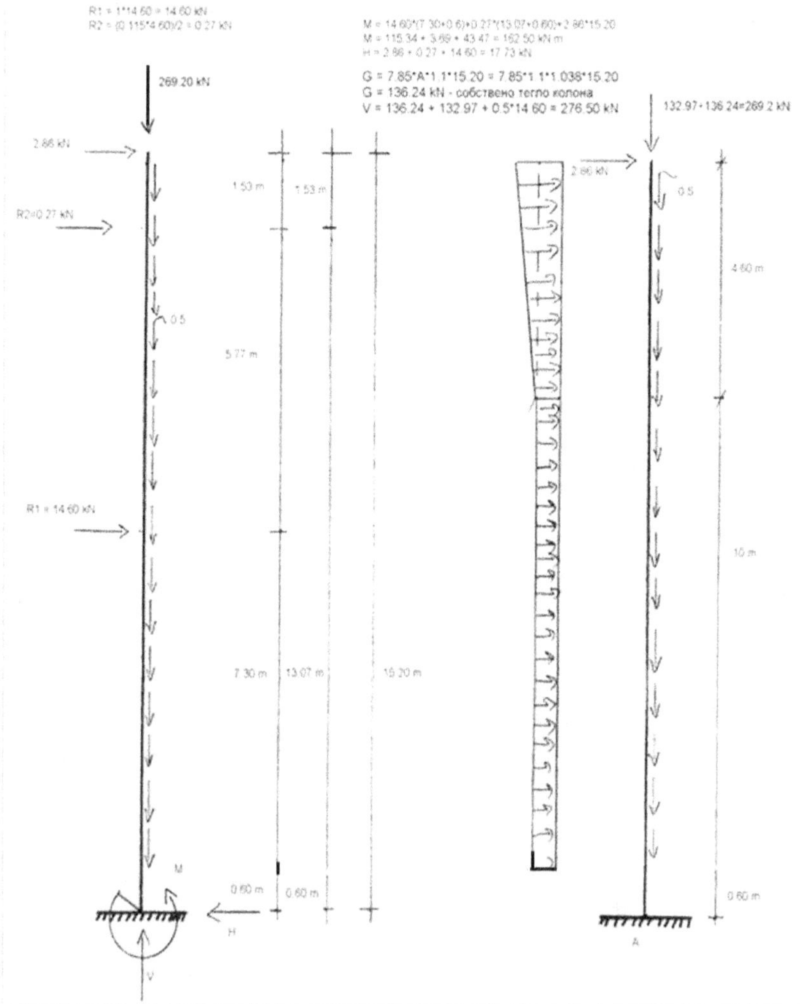

Figure 9.1 Static scheme, loads and calculations

9.2. FREE-BODY DIAGRAMS (FORCES DIGRAMS) – figure 9.2

9.3. DESIGN ON COLUMNS

► principal scheme for design – figure 9.3
► scheme of steel profile – figure 9.4
Calculation and design with the shaded part – the cross-section is not secured to stability.

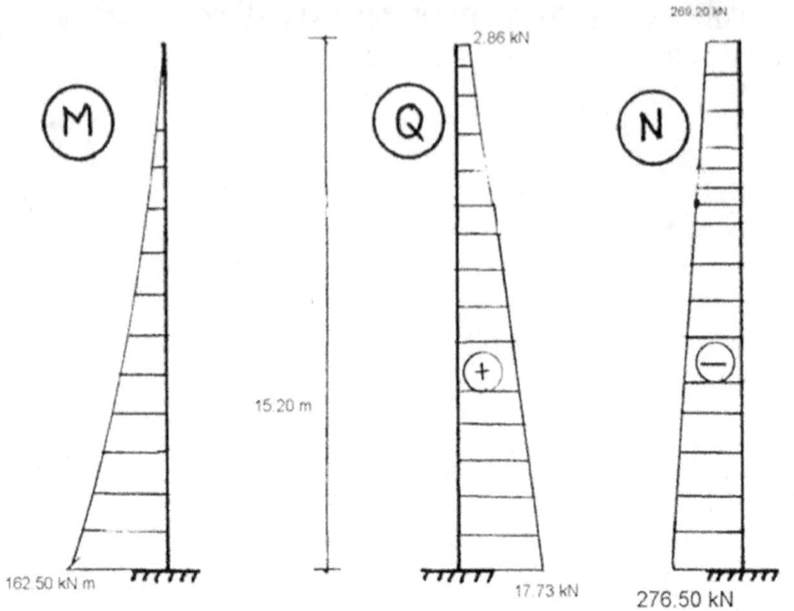

Figure 9.2 Free-body diagrams

Column design – N=276.50 kN, Q=17.73 kN, M=162.50 kN.m

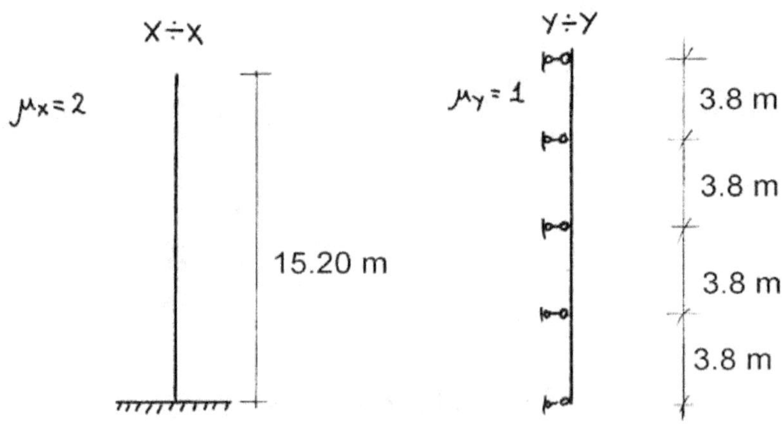

Figure 9.3 Principal scheme for design

$L_{ef,x} = \mu_x * l_{geom} = 2*15.20 = 30.40$ m $= 3040$ cm

$L_{ef,y} = \mu_y * l_{geom} = 1*3.80 = 3.80$ m $= 380$ cm

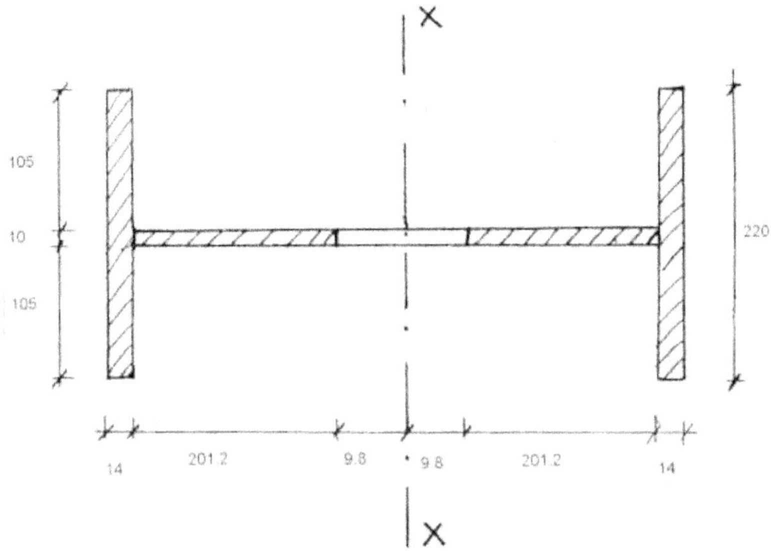

Figure 9.4 Scheme of steel profile

$$C_1 = 0.65 * t_w * \sqrt{\frac{E}{R_y}} = 0.65 * 1 * 30.96 = 20.12 \text{ cm - supporting width}$$

$$A = 1.4*22 + 1*42.2 + 1.4*22 - 1.96*1 = 101.84 \text{ cm}^2$$

$$\frac{b_f}{t_f} = \frac{220}{14} = 15.71 < 30$$

$$I_x = \frac{42.2^3 * 1}{12} + \frac{1.4^3 * 22 * 2}{12} + 44 * 1.4 * 21.8^2 - \frac{1.96^3 * 1}{12} = 35546.84 \text{ cm}^4$$

$$I_y = \frac{42.2 * 1^3}{12} + \frac{22^3 * 1.4 * 2}{12} - \frac{1^3 * 1.96}{12} = 2487.90 \text{ cm}^4;$$

$$W_x = \frac{I_x}{0.5 * h} = \frac{35546.84}{0.5 * 45} = 1579.86 \text{ cm}^3$$

$$i_x = \sqrt{\frac{I_x}{A}} = \sqrt{\frac{35546.84}{101.84}} = 18.68 \text{ cm} \; ; \; i_y = \sqrt{\frac{I_y}{A}} = \sqrt{\frac{2487.90}{101.84}} = 4.94 \text{ cm}$$

$$\frac{A_f}{A_w} = \frac{30.8}{40.24} = 0.77$$

$$\lambda_x = \frac{l_{ef,x}}{i_x} = \frac{3040}{18.68} = 162.74 < \lambda_u = 180$$

$$\lambda'_x = \lambda_x * \sqrt{\frac{R_y}{E}} = 162.74 * 0.0323 = 5.26$$

$$p_x = \frac{W_x}{A} = \frac{1579.86}{101.84} = 15.51 \text{ cm};$$

$$e_x = \frac{M}{N} = \frac{162.50}{276.50} = 0.5877 \text{ m} = 58.77 \text{ cm}$$

$$m_x = \frac{e_x}{p_x} = \frac{58.77}{15.51} = 3.79 \longrightarrow \eta = 1.25$$

$$m_{ef} = \eta * m_x = 1.25 * 3.79 = 4.74 \longrightarrow \phi_e = 0.133$$

$$\sigma = \frac{N}{\phi_e * A} = \frac{276.50}{0.133 * 101.84} = 20.41\ \mathrm{kN/cm^2} < \gamma_c * R_y = 21.5\ \mathrm{kN/cm^2}$$

► out-of-plane frame stability check (Y-Y)

$$\lambda_y = \frac{l_{ef,y}}{i_y} = \frac{380}{4.94} = 76.92 < \lambda_u = 180 \longrightarrow \phi_y = 0.762 ;$$

$$M_{ef} = \frac{2}{3} * M_{max} = \frac{2 * 162.5}{3} = 108.33\ \mathrm{kN.m}$$

$$p_x = 15.51\ \mathrm{cm} \quad ; \quad e'_x = \frac{M_{ef}}{N} = \frac{108.33}{276.50} = 0.3918\ \mathrm{m} = 39.18\ \mathrm{cm} ;$$

$$m_x = \frac{e'_x}{p_x} = \frac{39.18}{15.51} = 2.53 \ ; \ \beta = 1$$

$$\alpha = 0.65 + 0.05 * m_x = 0.65 + 0.05 * 2.53 = 0.777 \ ;$$

$$c = \frac{\beta}{(1+\alpha) * m_x} = \frac{1}{(1+0.777) * 2.53} = 0.222$$

$$\sigma = \frac{N}{\phi_y * c * A} = \frac{276.50}{0.222 * 0.762 * 101.84} = 16.05\ \mathrm{kN/cm^2} < 21.50\ \mathrm{kN/cm^2}$$

► checking local resistance of the plates

$$\frac{b_{ef}}{t_f} = \frac{10.5}{1.4} = 7.5 \le (0.36 + 0.1 * \lambda) * \sqrt{\frac{E}{R_y}} = (0.36 + 0.1 * 4) * 30.96 = 23.53$$

9.4. DESIGN ON BASE TO COLUMNS

► stability plate

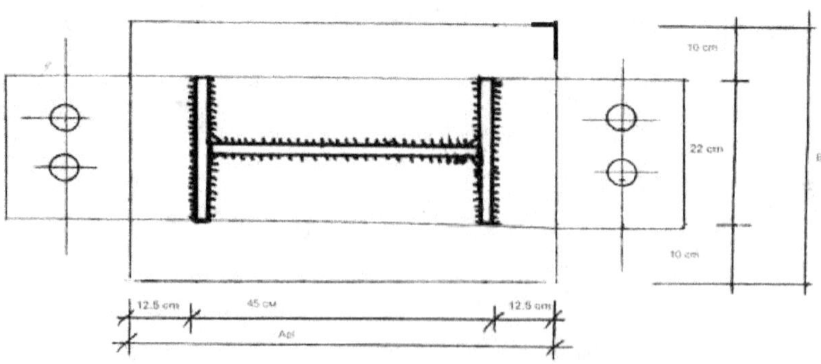

Figure 9.5 Scheme of stability plate

$a_{pl} = 45 + 2*12.5 = 70$ cm; $b_{pl} = 22 + 2*10 = 42$ cm; $R_b = 0.6$ kN/cm^2

$A_{loc} = (c+2*c_2)*(b_{pl}+2*c_1) = (41.94 + 2*28.06)*(42 + 10) = 5099.12$ cm^2

$A = c*b_{pl} = 41.94*42 = 1761.48$ cm^2

$$\sigma_1 = -\frac{N}{a_{pl} * b_{pl}} - \frac{6 * M}{b_{pl} * a_{pl}^2} = -\frac{276.50}{70 * 42} - \frac{6 * 16250}{42 * 70^2} = -0.568 \ \text{kN/cm}^2$$

$$\sigma_2 = -\frac{N}{a_{pl} * b_{pl}} + \frac{6 * M}{b_{pl} * a_{pl}^2} = -\frac{276.50}{70 * 42} + \frac{6 * 16250}{42 * 70^2} = +0.380 \ \text{kN/cm}^2$$

$$\phi_b = \sqrt[3]{\frac{A_{loc}}{A}} = \sqrt[3]{\frac{5099.12}{1761.48}} = 1.43 < 2.5;$$

$R_{b,loc} = \alpha * \phi_b * R_b = 1 * 1.43 * 0.6 = 0.858 \ \text{kN/cm}^2$

$\psi * R_{b,loc} = 0.75 * 0.858 = 0.644 \ \text{kN/cm}^2$

it follows that the support plate remains unchanged, i.e.

$a_{pl} = 45 + 2*12.5 = 70$ cm; $b_{pl} = 22 + 2*10 = 42$ cm

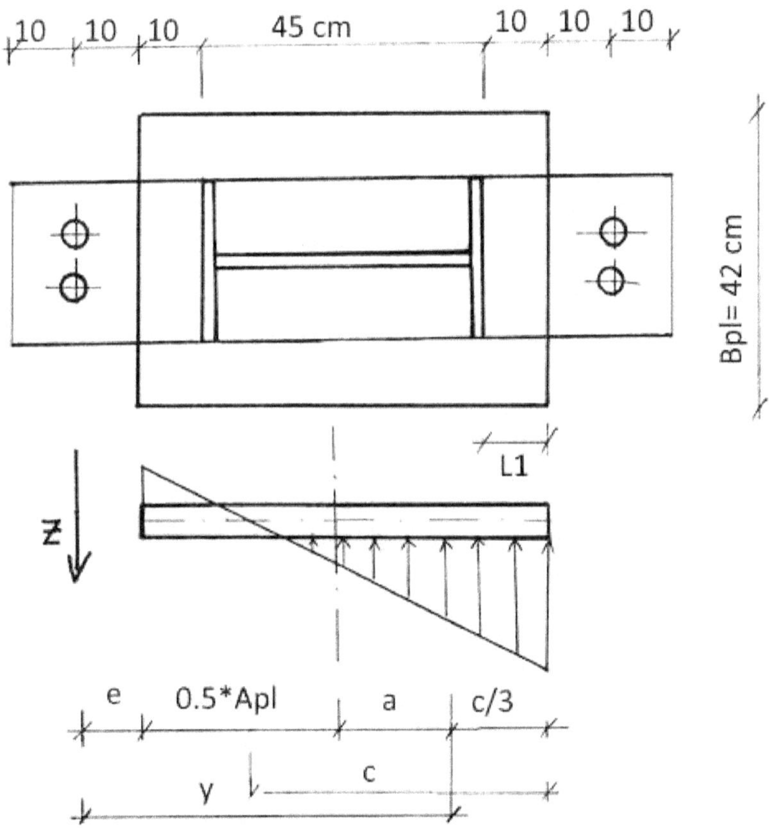

Figure 9.6 Scheme of plate

► determining the thickness of the base plate

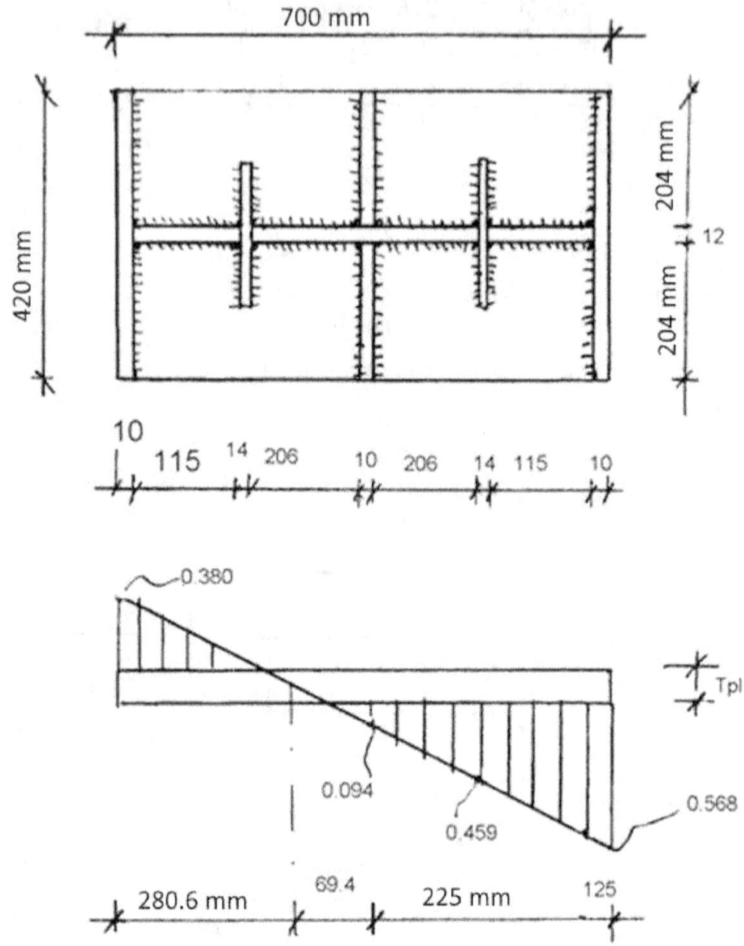

Figure 9.7 Scheme of base and dimensions

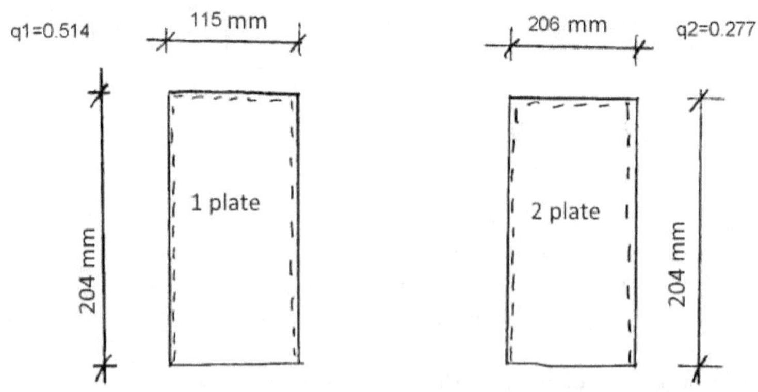

Figure 9.8 Scheme of static of plates and dimensions

For the sake of safety – we consider two fields, supported on 3 ends, but with different sizes. The load corresponds to the average MAX value of the corresponding field.

$$q_1 = \frac{0.568 + 0.459}{2} = 0.514 \quad ; \; q_2 = \frac{0.459 + 0.094}{2} = 0.277$$

For 1 plate: $M = \beta * q * a_1$; $a_1 = 11.5$ cm, $b_1 = 20.4$ cm,

$$\frac{b_1}{a_1} = \frac{20.4}{11.5} = 1.78 \longrightarrow \beta = 0.1298 \approx 0.130;$$

$$M = 0.130 * 0.514 * 11.5^2 = 8.84 \text{ kN.cm}$$

$T_{pl} \geq \sqrt{\frac{6*M}{R_y}} = \sqrt{\frac{6*8.84}{20.5}} = 1.61$ cm - it follows that it is accepted $T_{pl} = 20$ mm

For 2 plate: $M = \beta * q * a_1$; $a_1 = 20.6$ cm, $b_1 = 20.4$ cm,

$$\frac{b_1}{a_1} = \frac{20.4}{20.6} = 0.995 \approx 1 \longrightarrow \beta = 0.112; M = 0.112 * 0.277 * 20.6^2 = 13.16 \text{ kN.cm}$$

$T_{pl} \geq \sqrt{\frac{6*M}{R_y}} = \sqrt{\frac{6*13.16}{20.5}} = 1.96$ cm - it follows that it is accepted $T_{pl} = 20$ mm

► dimensioning of foundation bolts – they are dimensioned for force Z

$$a = \frac{a_{pl}}{2} - \frac{c}{3} = \frac{70}{2} - \frac{41.94}{3} = 21.02 \text{ cm} \;;$$

$$y = a_{pl} - \frac{c}{3} + e = 70 - \frac{41.94}{3} + 10 = 66.02 \text{ cm}$$

$$Z = \frac{M - N * a}{n * y} = \frac{162.50 * 100 - 276.50 * 21.02}{2 * 66.02} = 79.05 \text{ kN}$$

selected foundation bolt with diameter d=36 mm and strength R=106 kN according to scheme (2 plate) of ВСт3кп, welds made of electrodes type E46 with welding rod $k_f = 8$ mm

$$N_{1w} = \frac{N}{2 * n} + \frac{M}{h * n} = \frac{276.50}{2 * 4} + \frac{162.50 * 100}{45 * 4} = 124.84 \text{ kN}$$

$$l_w = \frac{N_{1w}}{\beta * k_f * \gamma_c * R_{wy}} = \frac{124.84}{0.7 * 0.8 * 1 * 18} = 12.40 \text{ cm}$$

Need to accept the crossbar with the construction height $h_{tr} = 60$ cm and $t_{tr} = 1.2$ cm.

$A_{tr} = 0.6 * 0.12 = 0.072 \text{ m}^2$; $W_{tr} = \frac{0.6*0.6*0.12}{6} = 0.0072 \text{ m}^3$

$$M_{tr}^{op} = \frac{\sigma_{max} * b_1 * l_1^2}{2} = \frac{0.568 * 42 * 12.5^2}{2} = 1863.75 \text{ kN.cm} = 18.64 \text{ kN.m}$$

$$Q_{tr}{}^{op} = \sigma_{max} * b_1 * l_1 = 0.568 * 42 * 12.5 = 298.20 \text{ kN}$$

$$M_{tr} = c_1 * Z = 0.125 * 79.05 = 9.88 \text{ kN.m} \quad ; \quad Q_{tr} = Z = 79.05 \text{ kN}$$

$$\sigma_{tr} = \frac{M_{tr}}{W_{tr}} = \frac{9.88}{0.0072} = 13.72 \text{ kN/cm}^2 \quad ; \quad \tau_{tr} = \frac{Q_{tr}}{A_{tr}} = \frac{79.05}{72} = 1.10 \text{ kN/cm}^2$$

the required leg of the welds, attaching the cross member to the support plate is:

$q_{tr} = \sigma_1 * b_1 * l_1 = 0.568*42*12.5 = 298.20 \text{ kN/cm}$

$$k_f = \frac{q_{tr}}{\beta_f * l_w * \gamma_f * R_{wy}} = \frac{298.20}{0.7 * 25 * 1 * 18} = 0.95 \text{ cm}$$

It follows that need to accept the welding rod $k_f = 10$ mm

9.5. FOUNDATION

► loads

$$N^n = \frac{N_c}{1.2} = \frac{276.50}{1.2} = 230.42 \text{ kN} \; ; \; Q^n = \frac{Q_c}{1.2} = \frac{17.73}{1.2} = 14.78 \text{ kN} \; ;$$

$$M^n = \frac{M_c}{1.2} = \frac{162.50}{1.2} = 135.42 \text{ kN.m}$$

$G_f^n = 0.85*a_f*b_f*\gamma_b*h = 0.85*1.80*23*a_f*b_f$;

γ_b=23 – Need using a lower grade of concrete than B15

► determining the main dimensions

$N_f^n = N^n + G_f^n = 230.42 + 0.85 * 1.80 * 23 * b_f * a_f =$

$= 230.42 + 35.19 * 1.4 * 2.8 = 368.36 \text{ kN}$

rectangular foundation with sides a_f = 280 cm ; b_f = 140 cm

$A = a_f * b_f = 280 * 140 = 39200 \text{ cm}^2$;

$$W = \frac{a_f^2 * b_f}{6} = \frac{280 * 280 * 140}{6} = 1829333.30 \text{ cm}^3$$

► determination of stresses

$$\sigma_{cp} = \frac{N_f^n}{A} = \frac{368.36}{39200} = 0.0094 \text{ kN/cm}^2$$

$$\sigma_1 = \frac{N_f^n}{A}\frac{M + Q * h}{W} = \frac{368.36}{39200} + \frac{13542 + 147.8 * 18}{1829333.30} = 0.0183 \text{ kN/cm}^2$$

$$\sigma_2 = \frac{N_f^n}{A} - \frac{M + Q * h}{W} = \frac{368.36}{39200} - \frac{13542 + 147.8 * 18}{1829333.30} = 0.000 \text{ kN/cm}^2$$

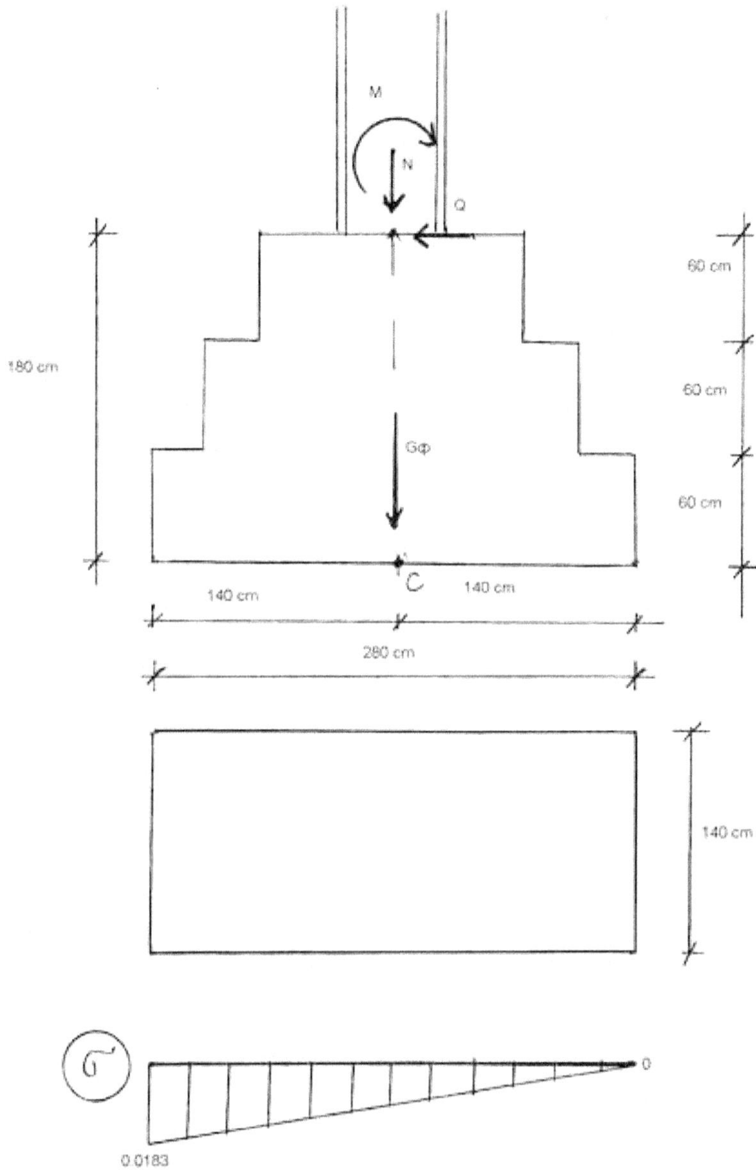

Figure 9.9 Scheme of foundation and dimensions

10. COLUMN AND FUNDATION ON "D" AXIS BY "Y"

10.1. STATIC SCHEME AND LOADS – figure 10.1

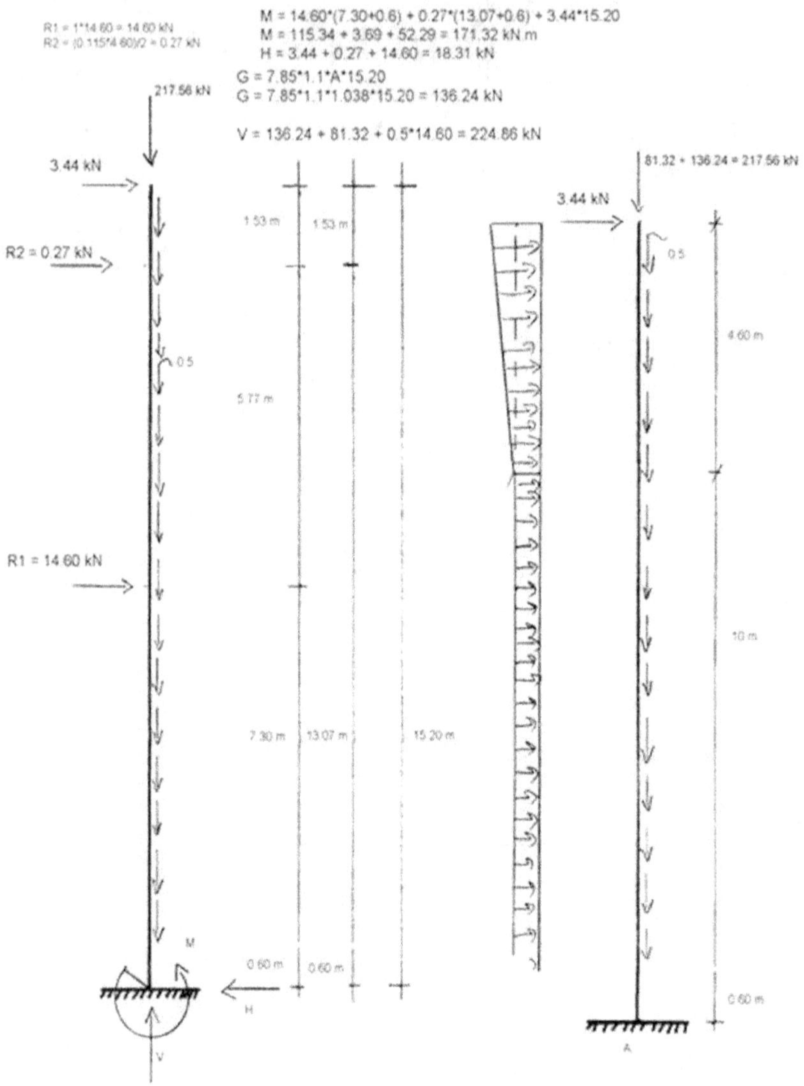

Figure 10.1 Static scheme, loads and calculations

9.2. FREE-BODY DIAGRAMS (FORCES DIGRAMS) – figure 10.2

9.3. DESIGN ON COLUMNS

- principal scheme for design – figure 10.3
- scheme of steel profile – figure 10.4

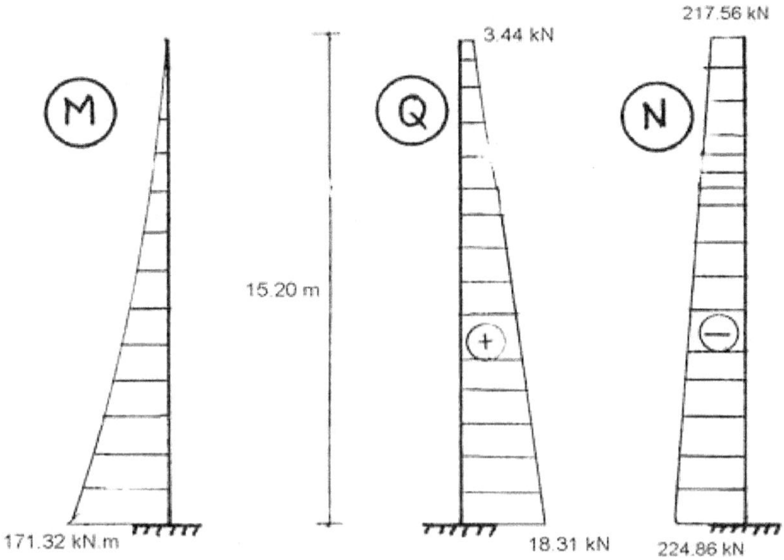

Figure 10.2 Free-body diagrams

Column design – N=224.86 kN, Q=18.31 kN, M=171.32 kN.m

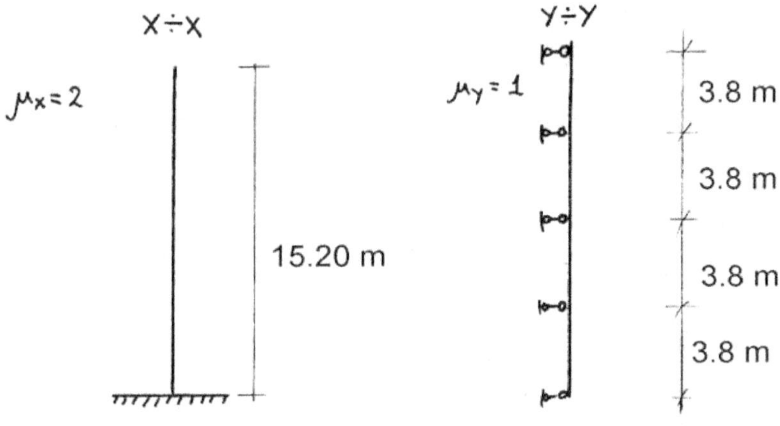

Figure 10.3 Principal scheme for design

$L_{ef,x} = \mu_x * l_{geom} = 2*15.20 = 30.40\ m = 3040\ cm$

$L_{ef,y} = \mu_y * l_{geom} = 1*3.80 = 3.80\ m = 380\ cm$

Calculation and design with the shaded part – the cross-section is not secured to stability.

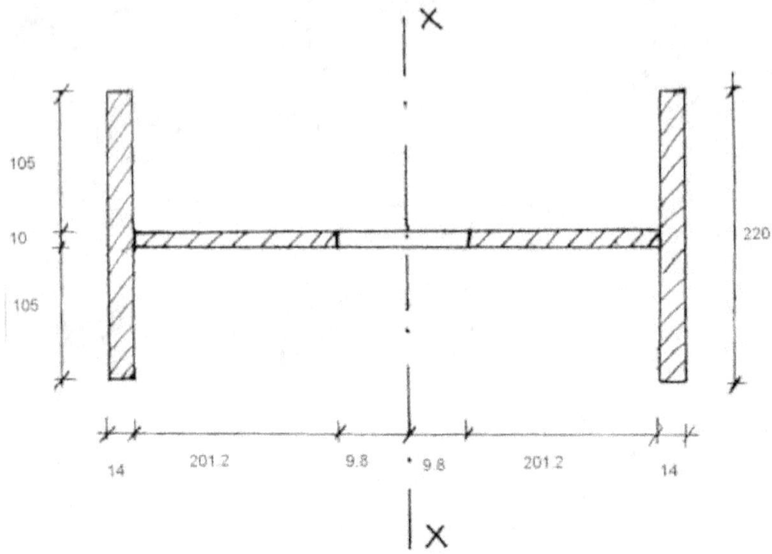

Figure 10.4 Scheme of steel profile

$$C_1 = 0.65 * t_w * \sqrt{\frac{E}{R_y}} = 0.65 * 1 * 30.96 = 20.12 \text{ cm - supporting width}$$

$$A = 1.4*22 + 1*42.2 + 1.4*22 - 1.96*1 = 101.84 \text{ cm}^2$$

$$\frac{b_f}{t_f} = \frac{220}{14} = 15.71 < 30$$

$$I_x = \frac{42.2^3 * 1}{12} + \frac{1.4^3 * 22 * 2}{12} + 44 * 1.4 * 21.8^2 - \frac{1.96^3 * 1}{12} = 35546.84 \text{ cm}^4$$

$$I_y = \frac{42.2 * 1^3}{12} + \frac{22^3 * 1.4 * 2}{12} - \frac{1^3 * 1.96}{12} = 2487.90 \text{ cm}^4;$$

$$W_x = \frac{I_x}{0.5 * h} = \frac{35546.84}{0.5 * 45} = 1579.86 \text{ cm}^3$$

$$i_x = \sqrt{\frac{I_x}{A}} = \sqrt{\frac{35546.84}{101.84}} = 18.68 \text{ cm} \; ; \; i_y = \sqrt{\frac{I_y}{A}} = \sqrt{\frac{2487.90}{101.84}} = 4.94 \text{ cm}$$

$$\frac{A_f}{A_w} = \frac{30.8}{40.24} = 0.77$$

$$\lambda_x = \frac{l_{ef,x}}{i_x} = \frac{3040}{18.68} = 162.74 < \lambda_u = 180$$

$$\lambda'_x = \lambda_x * \sqrt{\frac{R_y}{E}} = 162.74 * 0.0323 = 5.26$$

$$p_x = \frac{W_x}{A} = \frac{1579.86}{101.84} = 15.51 \text{ cm};$$

$a_{pl} = 45 + 2*12.5 = 70$ cm; $b_{pl} = 22 + 2*10 = 42$ cm; $R_b = 0.6$ kN/cm²

$A_{loc} = (c+2*c_2)*(b_{pl}+2*c_1) = (41.94 + 2*29.61)*(42 + 10) = 5179.72$ cm²

$A = c*b_{pl} = 40.39*42 = 1696.38$ cm²

$$\sigma_1 = -\frac{N}{a_{pl} * b_{pl}} - \frac{6 * M}{b_{pl} * a_{pl}^2} = -\frac{224.86}{70 * 42} - \frac{6 * 17132}{42 * 70^2} = -0.577 \ \text{kN/cm}^2$$

$$\sigma_2 = -\frac{N}{a_{pl} * b_{pl}} + \frac{6 * M}{b_{pl} * a_{pl}^2} = -\frac{224.86}{70 * 42} + \frac{6 * 17132}{42 * 70^2} = +0.423 \ \text{kN/cm}^2$$

$$\phi_b = \sqrt[3]{\frac{A_{loc}}{A}} = \sqrt[3]{\frac{5179.72}{1696.38}} = 1.45 < 2.5;$$

$R_{b,loc} = \alpha * \phi_b * R_b = 1 * 1.45 * 0.6 = 0.870 \ \text{kN/cm}^2$

$\psi * R_{b,loc} = 0.75 * 0.870 = 0.652 \ \text{kN/cm}^2$

it follows that the support plate remains unchanged, i.e.

$a_{pl} = 45 + 2*12.5 = 70$ cm; $b_{pl} = 22 + 2*10 = 42$ cm

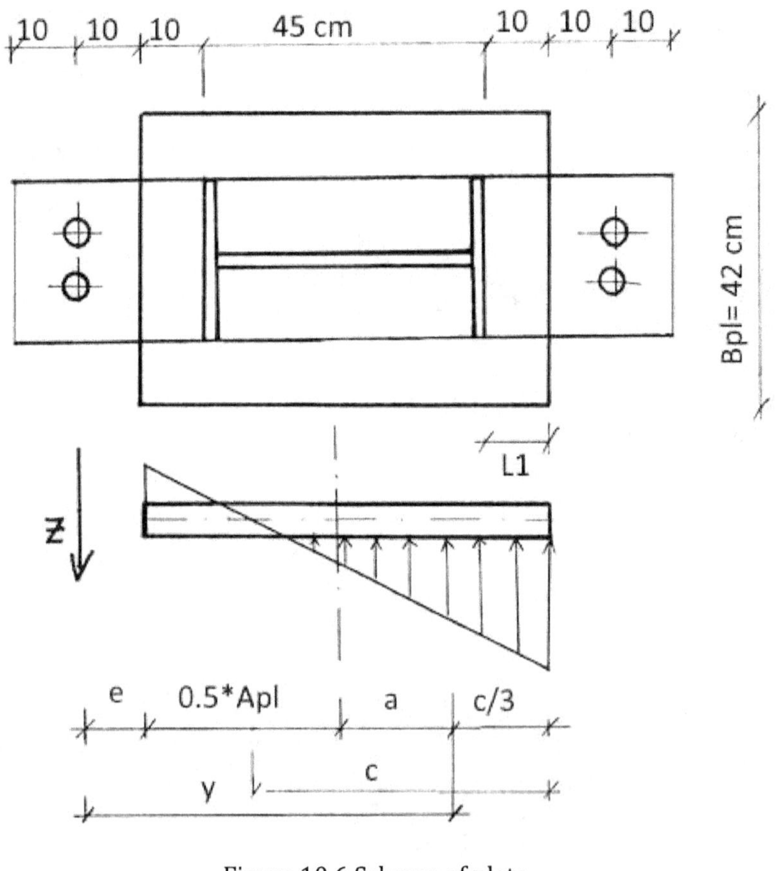

Figure 10.6 Scheme of plate

► determining the thickness of the base plate

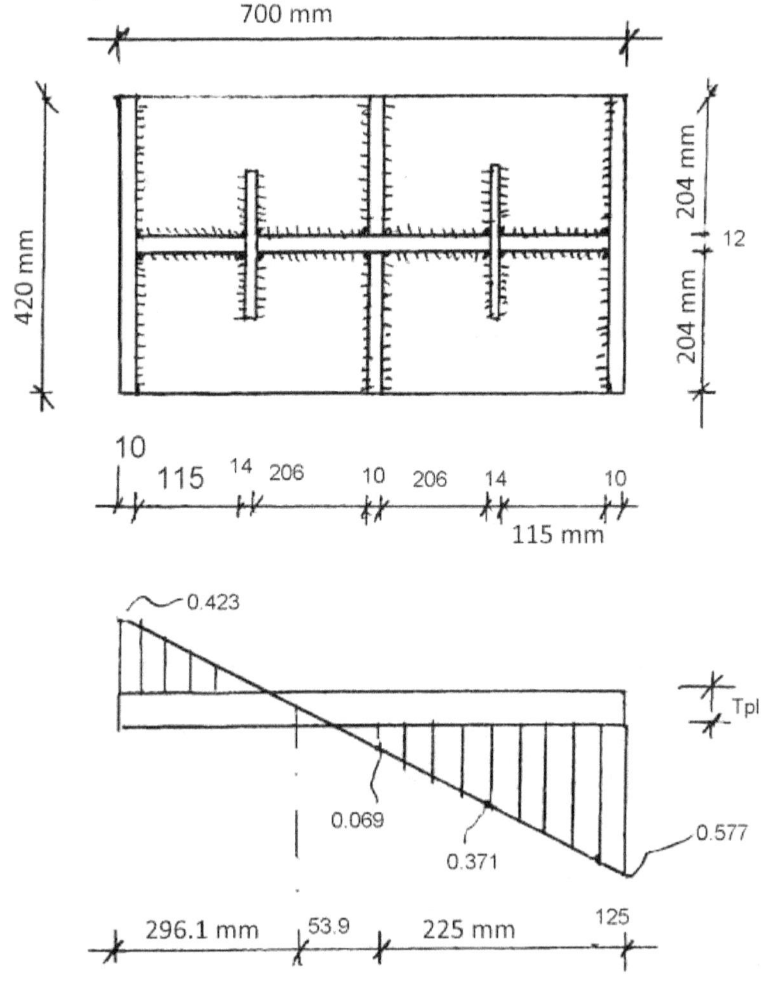

Figure 10.7 Scheme of base and dimensions

For the sake of safety – we consider two fields, supported on 3 ends, but with different sizes. The load corresponds to the average MAX value of the corresponding field.

$q_1 = \dfrac{0.577 + 0.371}{2} = 0.474$; $q_2 = \dfrac{0.371 + 0.069}{2} = 0.220$

For 1 plate: $M = \beta * q * a_1$; $a_1 = 11.5$ cm, $b_1 = 20.4$ cm,

$\dfrac{b_1}{a_1} = \dfrac{20.4}{11.5} = 1.78 \longrightarrow \beta = 0.1298 \approx 0.130$;

$M = 0.130 * 0.474 * 11.5^2 = 8.15$ kN.cm

$T_{pl} \geq \sqrt{\frac{6*M}{R_y}} = \sqrt{\frac{6*8.15}{20.5}} = 1.54$ cm - it follows that it is accepted $T_{pl} = 20$ mm

For 2 plate: $M = \beta * q * a_1$; $a_1 = 20.6$ cm, $b_1 = 20.4$ cm,

$\frac{b_1}{a_1} = \frac{20.4}{20.6} = 0.995 \approx 1 \rightarrow \beta = 0.112; M = 0.112*0.220*20.6^2 = 10.46$ kN.cm

$T_{pl} \geq \sqrt{\frac{6*M}{R_y}} = \sqrt{\frac{6*10.46}{20.5}} = 1.75$ cm - it follows that it is accepted $T_{pl} = 20$ mm

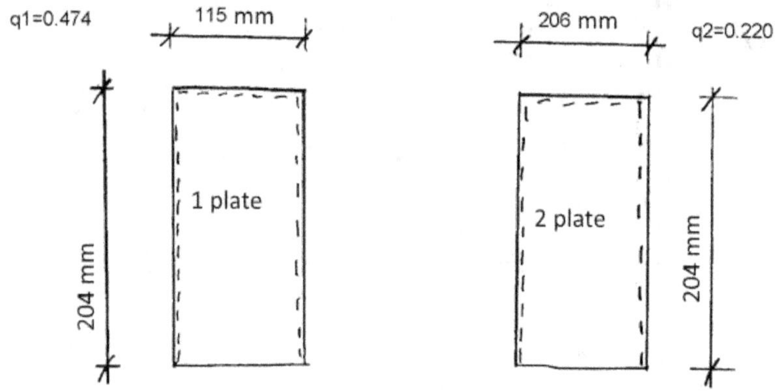

Figure 10.8 Scheme of static of plates and dimensions

► dimensioning of foundation bolts – they are dimensioned for force Z

$$a = \frac{a_{pl}}{2} - \frac{c}{3} = \frac{70}{2} - \frac{40.39}{3} = 21.54 \text{ cm} \;;$$

$$y = a_{pl} - \frac{c}{3} + e = 70 - \frac{40.39}{3} + 10 = 66.54 \text{ cm}$$

$$Z = \frac{M - N*a}{n*y} = \frac{171.32*100 - 224.86*21.54}{2*66.54} = 92.34 \text{ kN}$$

selected foundation bolt with diameter d=36 mm and strength R=106 kN according to scheme (2 plate) of ВСт3кп, welds made of electrodes type E46 with welding rod $k_f = 8$ mm

$$N_{1w} = \frac{N}{2*n} + \frac{M}{h*n} = \frac{224.86}{2*4} + \frac{171.32*100}{45*4} = 123.30 \text{ kN}$$

$$l_w = \frac{N_{1w}}{\beta * k_f * \gamma_c * R_{wy}} = \frac{123.30}{0.7*0.8*1*18} = 12.23 \text{ cm}$$

Need to accept the crossbar with the construction height $h_{tr} = 60$ cm and $t_{tr} = 1.2$ cm.

$A_{tr} = 0.6*0.12 = 0.072$ m^2 ; $W_{tr} = \frac{0.6*0.6*0.12}{6} = 0.0072$ m^3

$$M_{tr}^{op} = \frac{\sigma_{max} * b_1 * l_1^2}{2} = \frac{0.577*42*12.5^2}{2} = 1893.28 \text{ kN.cm} = 18.93 \text{ kN.m}$$

$$Q_{tr}{}^{op} = \sigma_{max} * b_1 * l_1 = 0.577*42*12.5 = 302.93 \text{ kN}$$

$M_{tr} = c_1 * Z = 0.125 * 92.34 = 11.54\ \text{kN.m}$; $Q_{tr} = Z = 92.34\ \text{kN}$

$\sigma_{tr} = \dfrac{M_{tr}}{W_{tr}} = \dfrac{11.54}{0.0072} = 16.03\ \text{kN/cm}^2$; $\tau_{tr} = \dfrac{Q_{tr}}{A_{tr}} = \dfrac{92.34}{72} = 1.28\ \text{kN/cm}^2$

the required leg of the welds, attaching the cross member to the support plate is:

$q_{tr} = \sigma_1 * b_1 * l_1 = 0.577*42*12.5 = 302.93\ \text{kN/cm}$

$$k_f = \frac{q_{tr}}{\beta_f * l_w * \gamma_f * R_{wy}} = \frac{302.93}{0.7 * 25 * 1 * 18} = 0.96\ \text{cm}$$

It follows that need to accept the welding rod $k_f = 10$ mm

9.5. FOUNDATION

► loads

$N^n = \dfrac{N_c}{1.2} = \dfrac{224.86}{1.2} = 187.38\ \text{kN}$; $Q^n = \dfrac{Q_c}{1.2} = \dfrac{18.31}{1.2} = 15.26\ \text{kN}$;

$M^n = \dfrac{M_c}{1.2} = \dfrac{171.32}{1.2} = 142.77\ \text{kN.m}$

$G_f^n = 0.85*a_f*b_f*\gamma_b*h = 0.85*1.80*23*a_f*b_f$;

γ_b=23 – Need using a lower grade of concrete than B15

► determining the main dimensions

$N_f^n = N^n + G_f^n = 187.38 + 0.85 * 1.80 * 23 * b_f * a_f =$

$= 230.42 + 35.19 * 1.4 * 2.8 = 325.33\ \text{kN}$

rectangular foundation with sides a_f= 280 cm ; b_f= 140 cm

$A = a_f * b_f = 280 * 140 = 39200\ \text{cm}^2$;

$$W = \frac{a_f^2 * b_f}{6} = \frac{280 * 280 * 140}{6} = 1829333.30\ \text{cm}^3$$

► determination of stresses

$$\sigma_{cp} = \frac{N_f^n}{A} = \frac{325.33}{39200} = 0.00830\ \text{kN/cm}^2$$

$$\sigma_1 = \frac{N_f^n}{A} \frac{M + Q * h}{W} = \frac{325.33}{39200} + \frac{14277 + 152.6 * 18}{1829333.30} = 0.0176\ \text{kN/cm}^2$$

$$\sigma_2 = \frac{N_f^n}{A} - \frac{M + Q * h}{W} = \frac{325.33}{39200} - \frac{14277 + 152.6 * 18}{1829333.30} = 0.000\ \text{kN/cm}^2$$

$M_{tr} = c_1 * Z = 0.125 * 92.34 = 11.54 \text{ kN.m}$; $Q_{tr} = Z = 92.34 \text{ kN}$

$\sigma_{tr} = \frac{M_{tr}}{W_{tr}} = \frac{11.54}{0.0072} = 16.03 \text{ kN/cm}^2$; $\tau_{tr} = \frac{Q_{tr}}{A_{tr}} = \frac{92.34}{72} = 1.28 \text{ kN/cm}^2$

the required leg of the welds, attaching the cross member to the support plate is:

$q_{tr} = \sigma_1 * b_1 * l_1 = 0.577*42*12.5 = 302.93 \text{ kN/cm}$

$$k_f = \frac{q_{tr}}{\beta_f * l_w * \gamma_f * R_{wy}} = \frac{302.93}{0.7 * 25 * 1 * 18} = 0.96 \text{ cm}$$

It follows that need to accept the welding rod $k_f = 10$ mm

9.5. FOUNDATION

► loads

$N^n = \frac{N_c}{1.2} = \frac{224.86}{1.2} = 187.38 \text{ kN}$; $Q^n = \frac{Q_c}{1.2} = \frac{18.31}{1.2} = 15.26 \text{ kN}$;

$M^n = \frac{M_c}{1.2} = \frac{171.32}{1.2} = 142.77 \text{ kN.m}$

$G_f^n = 0.85*a_f*b_f*\gamma_b*h = 0.85*1.80*23*a_f*b_f$;

γ_b=23 – Need using a lower grade of concrete than B15

► determining the main dimensions

$N_f^n = N^n + G_f^n = 187.38 + 0.85 * 1.80 * 23 * b_f * a_f =$

$= 230.42 + 35.19 * 1.4 * 2.8 = 325.33 \text{ kN}$

rectangular foundation with sides a_f = 280 cm ; b_f = 140 cm

$A = a_f * b_f = 280 * 140 = 39200 \text{ cm}^2$;

$$W = \frac{a_f^2 * b_f}{6} = \frac{280 * 280 * 140}{6} = 1829333.30 \text{ cm}^3$$

► determination of stresses

$$\sigma_{cp} = \frac{N_f^n}{A} = \frac{325.33}{39200} = 0.00830 \text{ kN/cm}^2$$

$$\sigma_1 = \frac{N_f^n}{A} \frac{M + Q * h}{W} = \frac{325.33}{39200} + \frac{14277 + 152.6 * 18}{1829333.30} = 0.0176 \text{ kN/cm}^2$$

$$\sigma_2 = \frac{N_f^n}{A} - \frac{M + Q * h}{W} = \frac{325.33}{39200} - \frac{14277 + 152.6 * 18}{1829333.30} = 0.000 \text{ kN/cm}^2$$

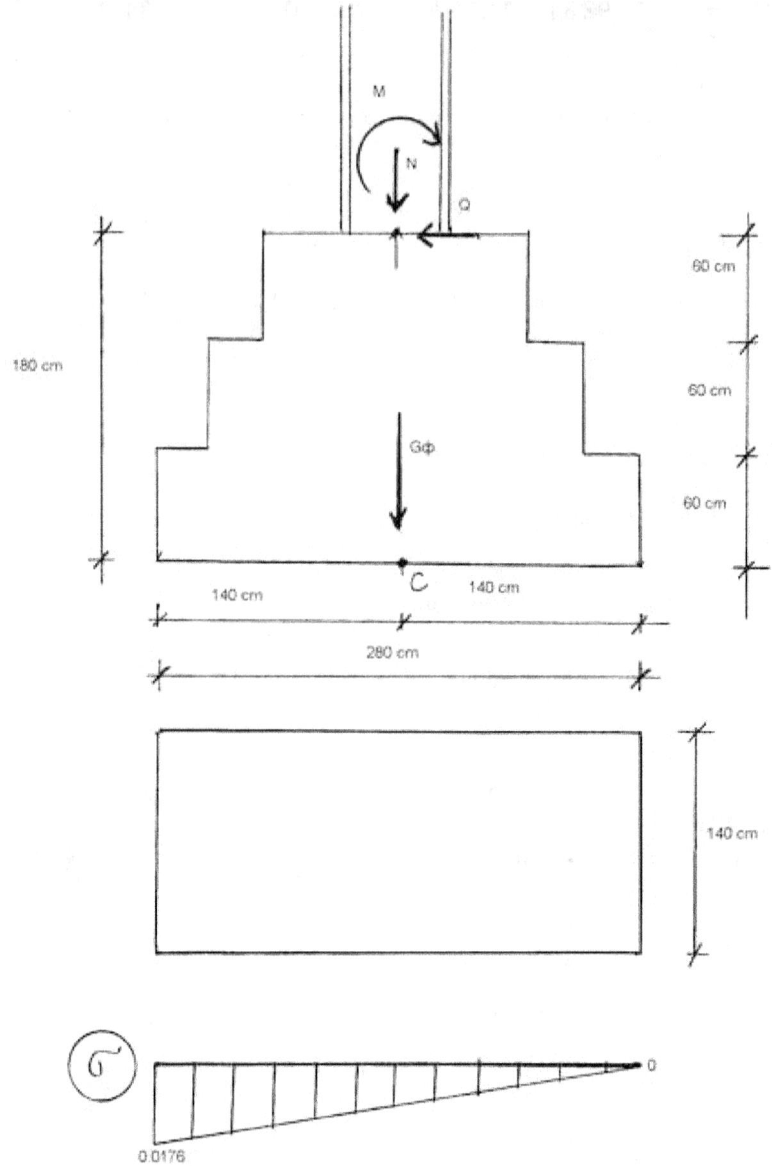

Figure 10.9 Scheme of foundation and dimensions

11. COLUMN AND FUNDATION ON "A1" AXIS – MAIN STRUCTURE

11.1. STATIC SCHEME AND LOADS – figure 11.1

11.3. DESIGN ON COLUMN

N=583.54 kN ,Q=32.78 kN , M=136.51 kN.m

► determination of μ frame action - figure 11.3

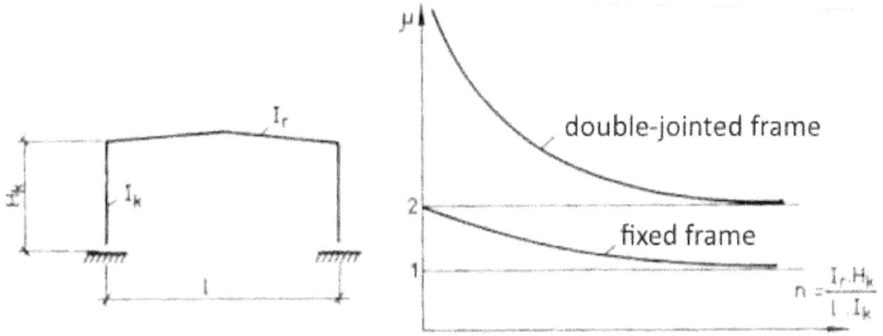

Figure 11.3 Determination of μ frame action

l = 36 метра , за I_r – average inertial axis of the truss, $h = \frac{2.60+4.20}{2} = 3.40$ m

$I_r = 4.31^2 * 65 + 65 * 170^2 + 3.07^2 * 31.2 + 31.2 * 170^2 = 2781681.50 \ \text{cm}^4$

H_k – column height to the average inertial axis (calculated column height – $l_{geom,x}$) ; $H_k = 15.20 - 2.60 + 3.40 = 16.00$ meter

► scheme of steel profile – figure 11.4

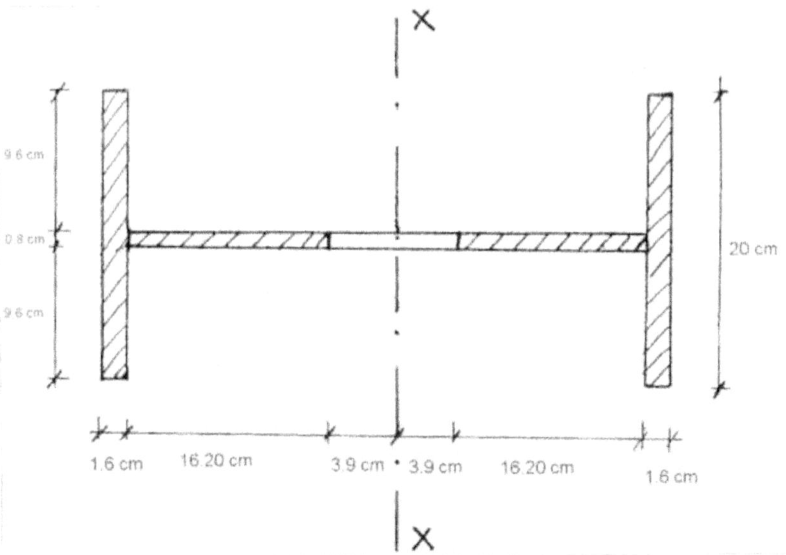

Figure 11.4 Scheme of steel profile

Calculation and design with the shaded part – the cross-section is not secured to stability.

$$C_1 = 0.65 * t_w * \sqrt{\frac{E}{R_y}} = 0.65 * 0.8 * 30.96 = 16.10 \text{ cm - supporting width}$$

$$A = 1.6*20 + 0.8*37.6 + 1.6*20 - 7.80*0.8 = 87.84 \text{ cm}^2$$

$$\frac{b_f}{t_f} = \frac{200}{16} = 12.50 < 30$$

$$I_x = \frac{37.6^3 * 0.8}{12} + \frac{1.6^3 * 20 * 2}{12} + 40 * 1.6 * 20.8^2 - \frac{7.8^3 * 0.8}{12} = 31217.15 \text{ cm}^4$$

$$I_y = \frac{37.6 * 0.8^3}{12} + \frac{20^3 * 1.6 * 2}{12} - \frac{0.8^3 * 7.80}{12} = 2134.60 \text{ cm}^4;$$

$$W_x = \frac{I_x}{0.5 * h} = \frac{31217.15}{0.5 * 40} = 1560.85 \text{ cm}^3$$

$$i_x = \sqrt{\frac{I_x}{A}} = \sqrt{\frac{31217.15}{87.84}} = 18.85 \text{ cm} \; ; \; i_y = \sqrt{\frac{I_y}{A}} = \sqrt{\frac{2134.60}{87.84}} = 4.93 \text{ cm}$$

$$\frac{A_f}{A_w} = \frac{32}{32.20} = 0.99 \approx 1$$

$$n = \frac{I_r * H_k}{l * I_k} = \frac{2781681.5 * 1600}{3600 * 31217.15} = 39.60 \, ;$$

$$\mu = \sqrt{\frac{n + 0.56}{n + 0.14}} = \sqrt{\frac{39.60 + 0.56}{39.60 + 0.14}} = 1.005 \approx 1$$

$$L_{ef,x} = \mu_x * l_{geom} = 1.01*16.00 = 16.16 \text{ m} = 1616 \text{ cm}$$

$$L_{ef,y} = \mu_y * l_{geom} = 1*3.80 = 3.80 \text{ m} = 380 \text{ cm}$$

$$\lambda_x = \frac{l_{ef,x}}{i_x} = \frac{1616}{18.85} = 85.73 < \lambda_u = 180$$

$$\lambda'_x = \lambda_x * \sqrt{\frac{R_y}{E}} = 85.73 * 0.0323 = 2.77$$

$$p_x = \frac{W_x}{A} = \frac{1560.85}{87.84} = 17.77 \text{ cm};$$

$$e_x = \frac{M}{N} = \frac{136.51}{538.54} = 0.2548 \text{ m} = 25.48 \text{ cm}$$

$$m_x = \frac{e_x}{p_x} = \frac{25.48}{17.77} = 1.43 \longrightarrow \eta = (1.90 - 0.1 * m_x) - 0.02 * (6 - m_x) * \lambda'_x$$

$$\eta = (1.90 - 0.1 * 1.43) - 0.02 * (6 - 1.43) * 2.77 = 1.50$$

$$m_{ef} = \eta * m_x = 1.50 * 1.43 = 2.15 \longrightarrow \phi_e = 0.330$$

$$\sigma = \frac{N}{\phi_e * A} = \frac{538.54}{0.330 * 87.84} = 18.60 \text{ kN/cm}^2 < \gamma_c * R_y = 21.5 \text{ kN/cm}^2$$

► out-of-plane frame stability check (Y-Y)

$$\lambda_y = \frac{l_{ef,y}}{i_y} = \frac{380}{4.93} = 77.08 \ < \ \lambda_u = 180 \longrightarrow \phi_y = 0.737 \ ;$$

$$M_{ef} = \frac{2}{3} * M_{max} = \frac{2 * 136.51}{3} = 91.00 \text{ kN.m}$$

$$p_x = 17.77 \text{ cm} \quad ; \ e'_x = \frac{M_{ef}}{N} = \frac{91.00}{538.54} = 0.1690 \text{ m} = 16.90 \text{ cm} \ ;$$

$$m_x = \frac{e'_x}{p_x} = \frac{16.90}{17.77} = 0.95 \ ; \ \beta = 1$$

$$\alpha = 0.65 + 0.05 * m_x = 0.65 + 0.05 * 0.95 = 0.698 \ ;$$

$$c = \frac{\beta}{(1+\alpha) * m_x} = \frac{1}{(1+0.698) * 0.95} = 0.620$$

$$\sigma = \frac{N}{\phi_y * c * A} = \frac{224.86}{0.620 * 0.737 * 87.84} = 13.41 \text{ kN/cm}^2 < 21.50 \text{ kN/cm}^2$$

► checking local resistance of the plates

$$\frac{b_{ef}}{t_f} = \frac{96}{16} = 6.0 \ \leq \ (0.36 + 0.1 * \lambda) * \sqrt{\frac{E}{R_y}} = (0.36 + 0.1 * 2.77) * 30.96 = 19.72$$

11.4. DESIGN ON BASE TO COLUMNS

► stability plate

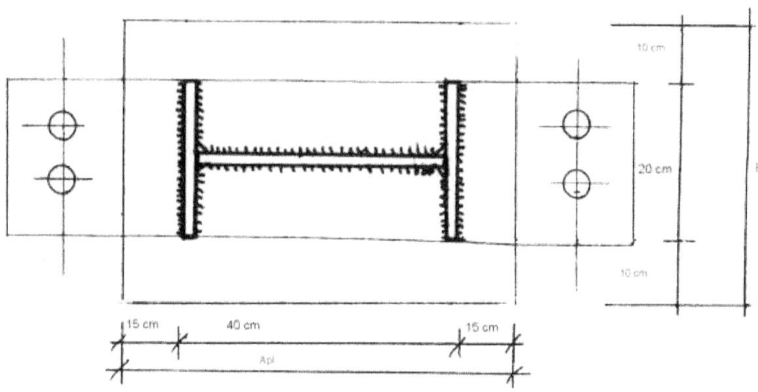

Figure 11.5 Scheme of stability plate

$a_{pl} = 45 + 2*15 = 70$ cm; $b_{pl} = 20 + 2*10 = 40$ cm; $R_b = 0.6$ kN/cm^2

$A_{loc} = (c + 2*c_2)*(b_{pl} + 2*c_1) = (54.31 + 2*15.69)*(40 + 10) = 5284.50$ cm^2

$A = c*b_{pl} = 54.31*40 = 2172.40$ cm^2

$$\sigma_1 = -\frac{N}{a_{pl} * b_{pl}} - \frac{6 * M}{b_{pl} * a_{pl}^2} = -\frac{538.54}{70 * 40} - \frac{6 * 13651}{40 * 70^2} = -0.540 \text{ kN/cm}^2$$

$$\sigma_2 = -\frac{N}{a_{pl} * b_{pl}} + \frac{6 * M}{b_{pl} * a_{pl}^2} = -\frac{538.54}{70 * 40} + \frac{6 * 13651}{40 * 70^2} = +0.226 \text{ kN/cm}^2$$

$$\phi_b = \sqrt[3]{\frac{A_{loc}}{A}} = \sqrt[3]{\frac{4284.50}{2172.40}} = 1.25 < 2.5\,;$$

$R_{b,loc} = \alpha * \phi_b * R_b = 1 * 1.25 * 0.6 = 0.750\ \text{kN/cm}^2$

$\psi * R_{b,loc} = 0.75 * 0.750 = 0.563\ \text{kN/cm}^2$

it follows that the support plate remains unchanged, i.e.

$a_{pl} = 40 + 2*15 = 70$ cm; $b_{pl} = 20 + 2*10 = 40$ cm

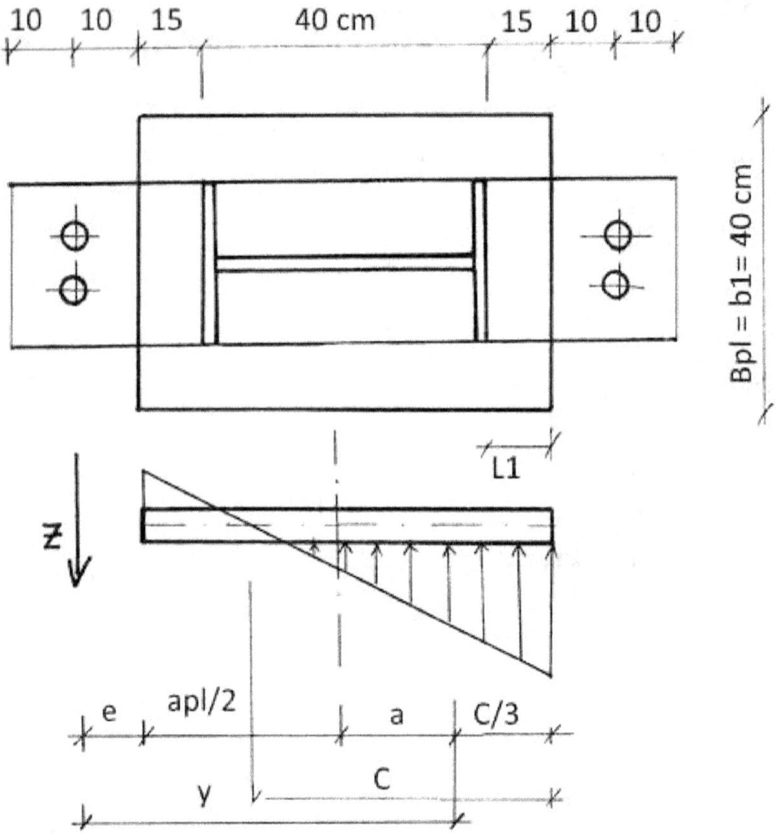

Figure 11.6 Scheme of plate

► determining the thickness of the base plate

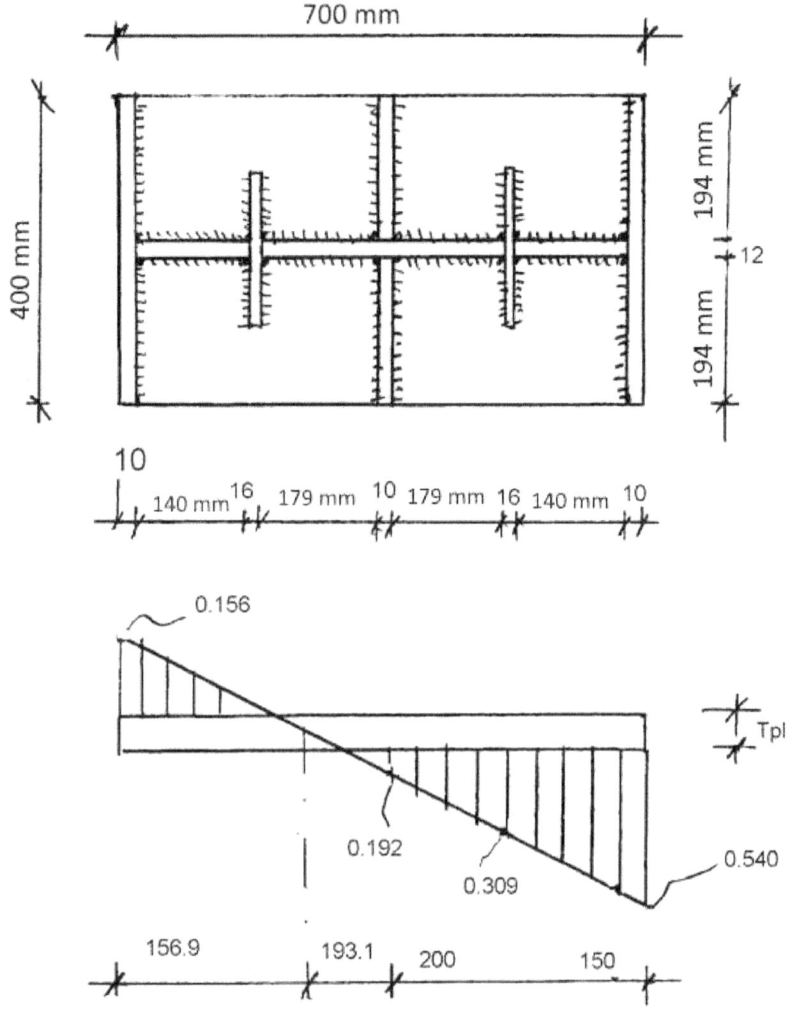

Figure 11.7 Scheme of base and dimensions

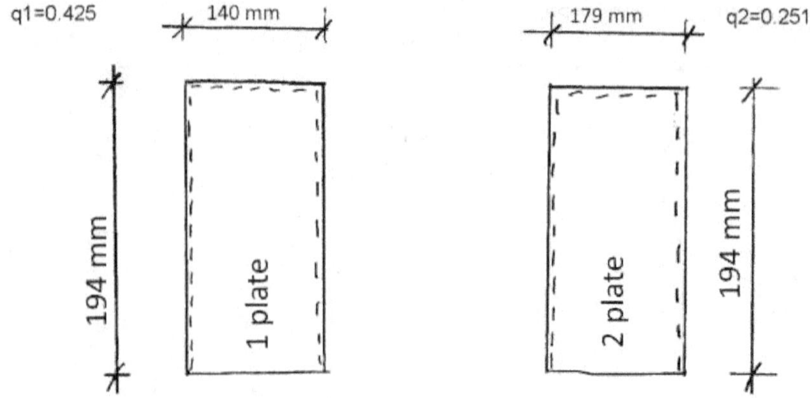

Figure 11.8 Scheme of static of plates and dimensions

For the sake of safety – we consider two fields, supported on 3 ends, but with different sizes. The load corresponds to the average MAX value of the corresponding field.

$$q_1 = \frac{0.540 + 0.309}{2} = 0.425 \quad ; \; q_2 = \frac{0.309 + 0.192}{2} = 0.251$$

For 1 plate: $M = \beta^*q^*a_1$; $a_1 = 14$ cm, $b_1 = 19.4$ cm,

$$\frac{b_1}{a_1} = \frac{19.4}{14} = 1.40 \longrightarrow \beta = 0.126;$$

$$M = 0.130^*0.425^*14^2 = 10.50 \text{ kN.cm}$$

$$T_{pl} \geq \sqrt{\frac{6*M}{R_y}} = \sqrt{\frac{6*10.50}{20.5}} = 1.75 \text{ cm}$$ - it follows that it is accepted $T_{pl} = 20$ mm

For 2 plate: $M = \beta^*q^*a_1$; $a_1 = 17.9$ cm, $b_1 = 19.4$ cm,

$$\frac{b_1}{a_1} = \frac{19.4}{17.9} = 1.005 \approx 1 \longrightarrow \beta = 0.112; M = 0.112^*0.251^*17.9^2 = 9.33 \text{ kN.cm}$$

$$T_{pl} \geq \sqrt{\frac{6*M}{R_y}} = \sqrt{\frac{6*9.33}{20.5}} = 1.65 \text{ cm}$$ - it follows that it is accepted $T_{pl} = 20$ mm

► dimensioning of foundation bolts – they are dimensioned for force Z

$$a = \frac{a_{pl}}{2} - \frac{c}{3} = \frac{70}{2} - \frac{54.31}{3} = 16.90 \text{ cm} \;;$$

$$y = a_{pl} - \frac{c}{3} + e = 70 - \frac{54.31}{3} + 10 = 61.90 \text{ cm}$$

$$Z = \frac{M - N * a}{n * y} = \frac{136.51 * 100 - 538.54 * 16.90}{2 * 61.90} = 36.75 \text{ kN}$$

selected foundation bolt with diameter d=36 mm and strength R=106 kN according to scheme (2 plate) of ВСт3кп, welds made of electrodes type E46 with welding rod $k_f = 8$ mm

$$N_{1w} = \frac{N}{2 * n} + \frac{M}{h * n} = \frac{538.54}{2 * 4} + \frac{136.51 * 100}{40 * 4} = 152.64 \text{ kN}$$

$$l_w = \frac{N_{1w}}{\beta * k_f * \gamma_c * R_{wy}} = \frac{152.64}{0.7 * 0.8 * 1 * 18} = 15.15 \text{ cm}$$

Need to accept the crossbar with the construction height $h_{tr} = 60$ cm and $t_{tr} = 1.2$ cm.

$A_{tr} = 0.6*0.12 = 0.072\ m^2$; $W_{tr} = \frac{0.6*0.6*0.12}{6} = 0.0072\ m^3$

$$M_{tr}^{op} = \frac{\sigma_{max} * b_1 * l_1^2}{2} = \frac{0.540 * 40 * 15^2}{2} = 2430 \text{ kN. cm} = 24.30 \text{ kN. m}$$

$$Q_{tr}{}^{op} = \sigma_{max} * b_1 * l_1 \quad = 0.540 * 40 * 15 = 324.00 \text{ kN}$$

$$M_{tr} = c_1 * Z = 0.125 * 36.75 = 5.51 \text{ kN.m} \quad ; \quad Q_{tr} = Z = 36.75 \text{ kN}$$

$$\sigma_{tr} = \frac{M_{tr}}{W_{tr}} = \frac{5.51}{0.0072} = 7.65 \text{ kN/cm}^2 \quad ; \quad \tau_{tr} = \frac{Q_{tr}}{A_{tr}} = \frac{36.75}{72} = 0.51 \text{ kN/cm}^2$$

the required leg of the welds, attaching the cross member to the support plate is:

$$q_{tr} = \sigma_1 * b_1 * l_1 = 0.540*40*15 = 324.00 \text{ kN/cm}$$

$$k_f = \frac{q_{tr}}{\beta_f * l_w * \gamma_f * R_{wy}} = \frac{324.00}{0.7 * 30 * 1 * 18} = 0.86 \text{ cm}$$

It follows that need to accept the welding rod $k_f = 10$ mm

9.5. FOUNDATION

► loads

$$N^n = \frac{N_c}{1.2} = \frac{583.54}{1.2} = 486.28 \text{ kN} \ ; \ Q^n = \frac{Q_c}{1.2} = \frac{32.78}{1.2} = 27.32 \text{ kN} \ ;$$

$$M^n = \frac{M_c}{1.2} = \frac{136.51}{1.2} = 113.76 \text{ kN.m}$$

$G_f^n = 0.85*a_f*b_f*\gamma_b*h = 0.85*1.80*23*a_f*b_f$;

γ_b=23 – Need using a lower grade of concrete than B15

► determining the main dimensions

$$N_f{}^n = N^n + G_f{}^n = 187.38 + 0.85 * 1.80 * 23 * b_f * a_f =$$
$$= 486.28 + 35.19 * 1.6 * 3.2 = 666.45 \text{ kN}$$

rectangular foundation with sides a_f= 320 cm ; b_f= 160 cm

$$A = a_f * b_f = 320 * 160 = 51200 \text{ cm}^2 \ ;$$

$$W = \frac{a_f{}^2 * b_f}{6} = \frac{320 * 320 * 160}{6} = 2730666.70 \text{ cm}^3$$

► determination of stresses

$$\sigma_{cp} = \frac{N_f{}^n}{A} = \frac{666.45}{51200} = 0.0130 \text{ kN/cm}^2$$

$$\sigma_1 = \frac{N_f{}^n}{A} \frac{M + Q * h}{W} = \frac{666.45}{51200} + \frac{11376 + 273.2 * 18}{2730666.70} = 0.0190 \text{ kN/cm}^2$$

$$\sigma_2 = \frac{N_f^n}{A} - \frac{M + Q * h}{W} = \frac{666.45}{51200} - \frac{11376 + 273.2 * 18}{2730666.70} = 0.007\ \mathrm{kN/cm^2}$$

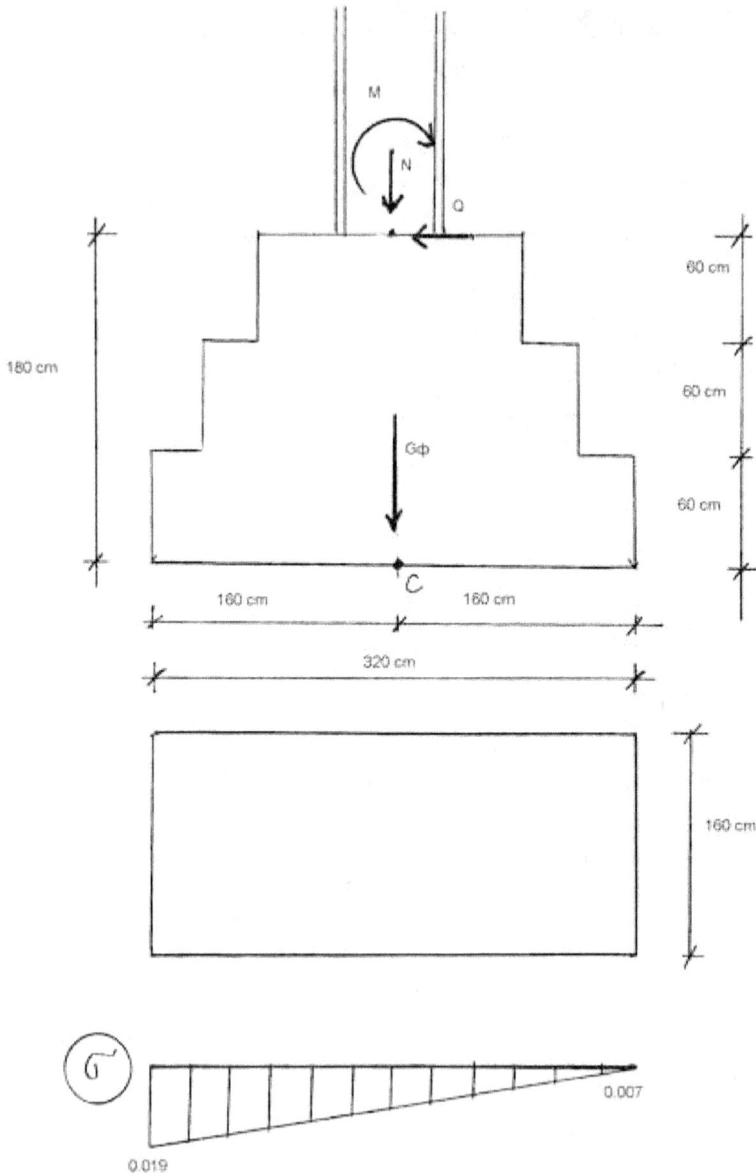

Figure 11.9 Scheme of foundation and dimensions

12. SPATIALLY STRENGTHENING

12.1. VERTICAL SPATIALLY STRENGTHENING

► static scheme and loads (figure 12.1)

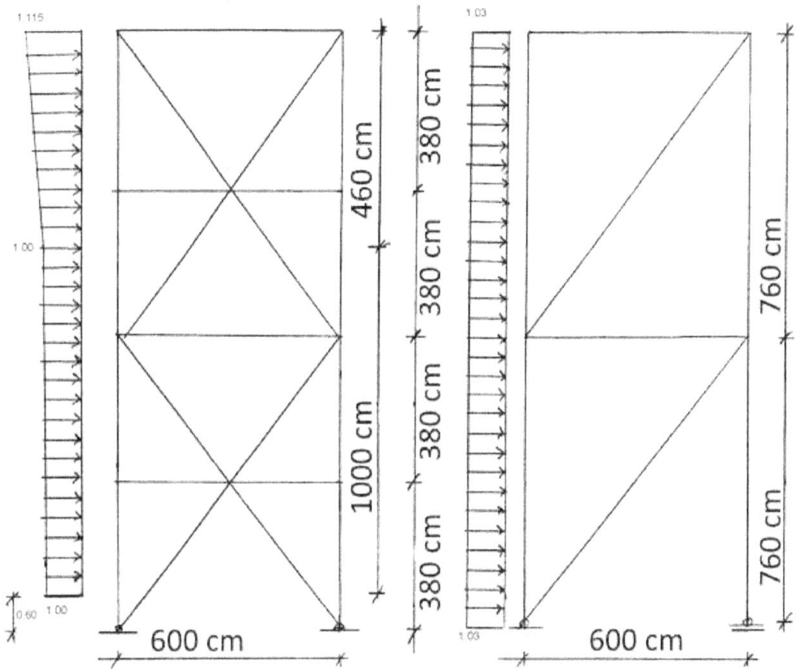

Figure 12.1 Static scheme and loads

► results – figure 12.2

► design

$D_1 = 18.80+30.50 = 49.30$ kN (tension)

$$A_{neobx} = \frac{D_1}{\gamma_c * R_y} = \frac{49.30}{0.95 * 22.5} = 2.30 \ cm^2$$

steel profile (cross – section) ⌋⌊ 70.70.5

$t_{pl} = 10$ mm , $A = 13.72$ cm^2 , $i_x = 2.16$ cm , $i_y = 3.23$ cm

$l_{ef,x} = 0.8 * l_{geom,x} = 0.8 * 355.11 = 284.09$ cm,

$l_{ef,y} = 1 * l_{geom,y} = 1 * 355.11 = 355.11$ cm

$$\lambda_x = \frac{284.09}{2.16} = 131.52 < \lambda_{gr} = 400 \ , \quad \lambda_y = \frac{355.1}{3.23} = 109.94 < \lambda_{gr} = 400$$

$$\sigma = \frac{D_1}{A} = \frac{49.30}{13.72} = 3.59 \ kN/cm^2$$

$V_1 = - 11.65 - 18.90 = - 30.55$ kN (compression)

$$A_{neobx} = \frac{V_1}{\phi * \gamma_c * R_y} = \frac{30.55}{0.6 * 0.8 * 22.5} = 2.83 \ cm^2$$

steel profile (cross – section) ⌋⌊ 100.100.10

$t_{pl} = 10 \text{ mm}\ ,\ A = 38.40 \text{ cm}^2\ ,\ i_x = 3.05 \text{ cm}\ ,\ i_y = 4.52 \text{ cm}$

$$l_{ef,x} = 600 \text{ cm},\ l_{ef,y} = 600 \text{ cm},\ \lambda_x = \frac{600}{3.05} = 196.72,\ \lambda_y = \frac{600}{4.52} = 132.74$$

$\lambda_{max} = 196.72 \longrightarrow \phi = 0.180$

$$\sigma = \frac{V_1}{\phi * A} = \frac{30.55}{0.180 * 38.40} = 4.42 \text{ kN/cm}^2$$

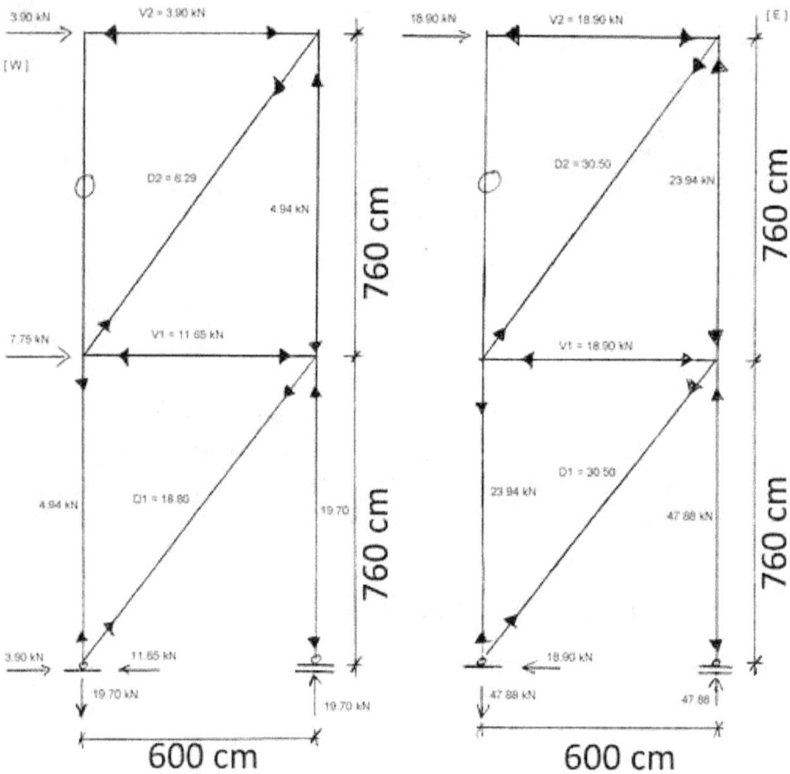

Figure 12.2 Results

12.2. HORIZONTAL SPATIALLY STRENGTHENING

► static scheme and loads (figure 12.1)

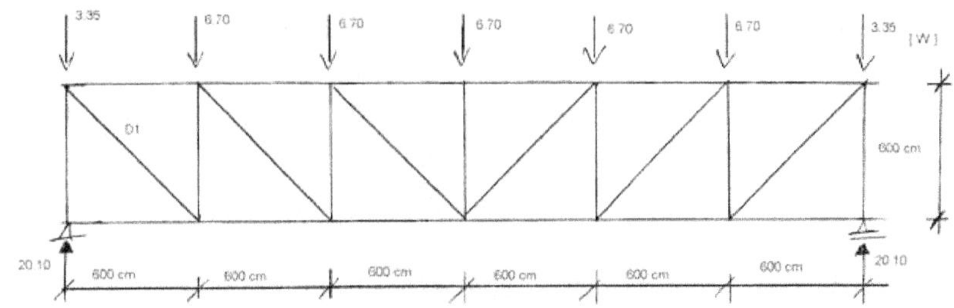

Figure 12.1 Static scheme and loads – normal case

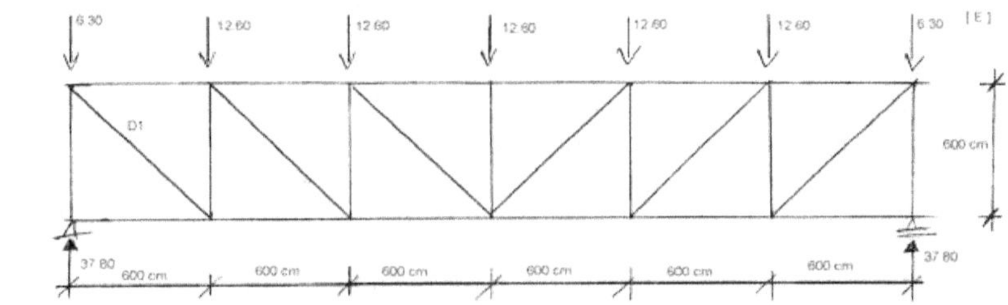

Figure 12.1 Static scheme and loads – earthquake case

► design - only diagonal (tension)

$\cos\alpha * D_1 + 3.35 = 20.10$ kN

$$D_1 = \frac{20.10 - 3.35}{\cos 45^0} = 23.70 \text{ kN};$$

$$l_{el,x} = \mu_x * l_{geom,x} = 0.8 * 424.26 = 339.41 \text{ cm}$$

$\cos\alpha * D_1 + 6.30 = 37.80$ kN (tension)

$$D_1 = \frac{37.80 - 6.30}{\cos 45^0} = 44.55 \text{ kN};$$

$$l_{el,y} = \mu_y * l_{geom,y} = 1 * 424.26 = 424.26 \text{ cm}$$

steel profile (cross − section) ⌋ ⌊ 70.70.5

$$t_{pl} = 10 \text{ mm}, A = 13.72 \text{ cm}^2, i_x = 2.16 \text{ cm}, \; i_y = 3.23 \text{ cm}$$

$$l_{ef,x} = 0.8 * l_{geom,x} = 0.8 * 424.26 = 339.41 \text{ cm},$$

$$l_{ef,y} = 1 * l_{geom,y} = 1 * 424.26 = 424.26 \text{ cm}$$

$$\lambda_x = \frac{339.41}{2.16} = 107.41 < \lambda_{gr} = 400, \quad \lambda_y = \frac{424.26}{3.23} = 131.35 < \lambda_{gr} = 400$$

$$\sigma = \frac{D_1}{A} = \frac{44.55 + 23.70}{13.72} = \frac{68.25}{13.72} = 4.98 \text{ kN/cm}^2$$

13. CHOICE OF CRANE FOR ASSEMBLY OF ELEMENTS

13.1. TECHNOLOGICAL ANALYSIS

The building is a one-story metal structure - single-span with dimensions in plan 36x42 meters. The building plan allows the full width of the building to be used for the implementation of the assembly processes. For the implementation of the assembly, it is necessary to implement temporary roads, to trace all axes in both directions and to provide all necessary material, technical and labor resources for the implementation of the assembly work.
The volume of the assembly work - installation of columns and trusses (sample technological scheme), with one crane.

13.2. DETERMINATION OF THE ASSEMBLY CRANE

The determination of the required crane is done by selecting the technical parameters. This selection is carried out in accordance with the basic installation schemes. The crane moves (for columns and trusses in an "X" pattern) inside the building, and when installing the trusses in an "Y" pattern outside the building. The installation crane is selected by comparing and maximizing the requirements for compliance between the installation and technical parameters of the elements and the technical and operational characteristics of the crane. The possibility of feeding the elements into the design position depends on the geometric limitation for entering the crane space of the point located on the most protruding edge of the element towards the boom [9].

$$\sqrt[3]{Lo^2} = \sqrt[3]{(x+a)^2} + \sqrt[3]{y^2}\ [m]\ ;\ \mathrm{Lo} = \frac{y}{\sin\alpha^0} + \frac{x+a}{\cos\alpha^0}\ [m]\ ;$$

$$\alpha^0 = \mathrm{arctg}\sqrt[3]{\frac{y}{x+a}}\ ;$$

$$\mathrm{Zo} = \mathrm{a} + \sqrt[3]{y^2 * (x+a)}\ [m]\ ;\ \mathrm{Bo} = \mathrm{x} + \mathrm{Zo} + \mathrm{C}\ [m]\ ;$$

$$\mathrm{Bo} = \mathrm{x} + \mathrm{a} + \frac{\mathrm{Bo}}{\mathrm{tg}\alpha o}\ [m]\ ;$$

$$(\mathrm{x}+\mathrm{a})^*\mathrm{y} = \mathrm{t}^*(\mathrm{z}\text{-}\mathrm{a})\ ;\ \mathrm{L} = \frac{\mathrm{x}+\mathrm{z}}{\mathrm{z}\text{-}\mathrm{a}} * \sqrt{(\mathrm{z}-\mathrm{a})^2 + \mathrm{y}^2}\ ;$$

$$\mathrm{t} = \mathrm{Ht} + \mathrm{Hn}\ [\mathrm{m}]\ ;\ \mathrm{y} = \mathrm{Hm} + \mathrm{d} + \mathrm{He} + \mathrm{Hmin}$$

where:
Lo – the minimum required length of the crane boom to fit the point into the under-beam space
L - boom length
X – installation width (horizontal projection of the boom onto the mounted building element)

Y – installation height (from the boom heel to the mounting horizon through the beam)
αo – the optimal angle of inclination of the boom corresponding to the minimum length
a ≥ 1.5 m – safe horizontal distance from the point to the boom
axis Z – overall distance from the boom heel to the mounted building element
B – installation range of the crane
t – distance from the horizontal through the point to the top
C – distance from the boom heel to the crane rotation axis
He – height and width in the suspended position
Hm – working height of the lifting device
Hn = 1.5 m – minimum height of the load-lifting pulley block in stowed position
Ho = 1.8 m – distance from the ground to the toe of the crane boom
Hk – required height to the crane hook
Hm – installation height, measured from the crane travel level to the design position
When comparing the characteristics of the crane, the following requirements must be met:
Lkr > Lo; Qkr > Qm = qe+qt;
Bmin < Bkr < B max Qm – mass of the assembly unit;
qe – mass of the element;
qt – mass of the rigging device;
Bmin and Bmax – min and max permissible range of action for assembly;

Table 13.1 Table for determination of characteristics of crane

	Column	Main truss	Truss by "Y" (42 meter)	Truss by "X" (36 meter)
qe, [t]	1451	4883	510	3702
He, [m]	15,20	4,20	0,14	2,60
m, [m]	0,45	0,30	0,30	0,26
qt, [t]	0,36	1,43	0,36	1,43
Ht, [m]	1,40	5,97	4,00	5,97
Qm, [t]	1452	4885	511	3704
Hm, [m]	0	14,60	14,60	14,60
Hk, [m]	16,60	25,80	19,75	24,20
X, [m]	0,225	0,15	18,00	0,13
Y, [m]	14,40	18,00	13,94	16,40
Lo, [m]	19,96	23,76	47,07	21,95

Zo, [m]	10,65	9,62	17,09	9,10
C1 = +1.50, [m]	12,38	11,27	36,59	10,73
Lkr, [m]	21,00	29,00	42,00	29,00
Z1, [m]	5,85	4,29	3,21	3,66
B1, [m]	6,08	4,44	5,37	3,79
Bmin, [m]	6,08	4,44	5,37	3,79
Bmax, [m]	19,70	12,5	30	23,03
Qmax, [t]	7,50	7	0,5	7,00
Qmin, [t]	4,50	2,50	0	2,50
Hk,max, [m]	17,5	27,9	34	27,90
Hk,min, [m]	16,6	25,6	29	25,60

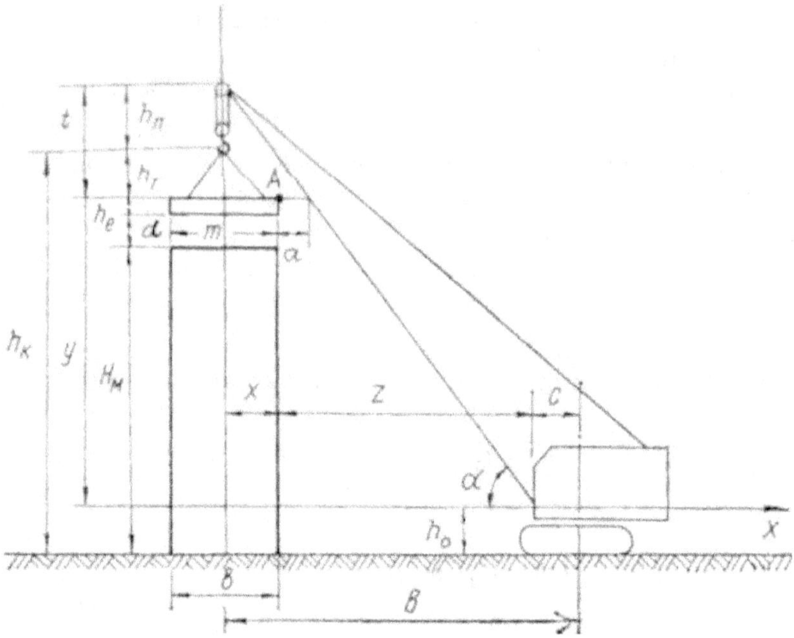

Figure 13.1 Main scheme for determination of characteristics of crane [7-9]

Based on these parameters, the MDK 504 crane was selected.

► technology of the installation process (method of implementation)

step1. Preparatory work - clearing and marking the site.

step 2. Earthworks - excavation for foundations, transport to the depot, etc.

step 3. Zero cycle - poured concrete for the foundations, formwork and demoulding of foundations, reinforcement and concreting of the foundations.

step 4. Installation "Columns" - all types of columns are installed,

reinforcement and concreting of the concrete pavement, installation of vertical windbreaks (spatial reinforcement of the columns)
step 5. Installation "Trusses" - the trusses are installed along "X" and "Y", as well as the horizontal spatial reinforcement.
step 6. Installation "Gantry truss" - the gantry truss is installed.
step 7. Installation "Elements" – the elements of the superstructure (roof), capitals, etc. are installed.
step 8. Finishing works – roof waterproofing and others.

► safety techniques

1. All workers must be instructed in the safe performance of installation works and must be equipped with personal protective equipment.
2. The serviceability of the MDK 504 crane must be certified by the inspection bodies. The stabilization of the outriggers in the installation positions must be carried out in accordance with the methodology.
3. The gripping attachments and installation traverses must be used according to their load capacity and tested periodically.
4. The elements are hung from the crane hook after they are strengthened. The necessary ladders, platforms, scaffolding and other technical equipment must be installed in advance where necessary.
5. It is prohibited for people to stand under the lifted loads.
6. All requirements in the documents must be observed:

⇒ Regulation on physical protection systems for construction sites.
⇒ Regulations on occupational safety in construction.
⇒ Regulation on the design and safety of the operation of lifting cranes.

14. CONCLUSION

The structural composition of a hangar (production building) offers real opportunities for polyvariance, both in terms of overall dimensions - through modularity, and in terms of loads and adaptability to various technical and technological requirements. It is clearly possible to conditionally "unify" real structural elements and schemes, offering real opportunities in the field of reconstructions and modernizations of modern buildings, offering a solution to the conflict between investment durability and functional (by production purpose) short-term durability of industrial premises, through a non-destructive method of reconstruction and modernization of the same. The latter actually equalizes the real "cost" after the first reconstruction of buildings made of metal structures with those made of reinforced concrete, as the "term" of reconstruction or "dead economic time" (the time during which the building cannot be used for its new purpose) is in favor of metal structures. The wide possibilities for variation of both the external forms - visible skeleton, and the variable internal distribution of the proposed constructive model determine the main economic indicators that make the proposed development competitive and attractive to investors. With this development of mine, I believe that I have managed to understand the two basic principles of the constructive-cognitive plan, namely: first - to doubt before being convinced and second - not to deny before trying to understand.

REFERENCES

[1] "Normi za proektirane na stomanenie konstrucii", ABC Technika, 1999 [in Bulgarian]
[2] "Natovarvaniya I vazdestviya varxu sgradi i saurajeniya", ABC Technika, 1999 [in Bulgarian]
[3] "Normi za proektirane na sgradi i saurajeniya v zemetrasni raioni", ABC Technika, 1999 [in Bulgarian]
[4] Chacov, T. "Stomanenie konstrucii", Technika 1978 [in Bulgarian]
[5] Brainov, M. and Venkov, L. "Stomanenie konstrucii", Technika 1991 [in Bulgarian]
[6] Venkov, L. and etc. "Stomanenie konstrucii - rakovodstvo za proektirane", UASG, 1998 [in Bulgarian]
[7] Petkov, Y. "Tehnologiya na stroitelnite procesi", Technika 1978 [in Bulgarian]
[8] Petkov, Y. , Zarkova, L.and Donchev, Ch. "Tehnologiya na stroitelnoto proizvodstvo, sbornik ot zadaci i primeri" Technika 1994 [in Bulgarian]
[9] "Montajni kranove v stroitelstvoto" Technika 1978 [in Bulgarian]
[10] Bankov, B. "Stroitelna statica - staticeski neopredelimi sistemi" UASG, 2002 [in Bulgarian]

www.ingramcontent.com/pod-product-compliance
Lightning Source LLC
LaVergne TN
LVHW022313200926
848481LV00004B/1323